DATA QUALITY CONTROL

STATISTICS: Textbooks and Monographs

Vol. 1: The Generalized Jackknife Statistic, *H. L. Gray and W. R. Schucany*
Vol. 2: Multivariate Analysis, *Anant M. Kshirsagar*
Vol. 3: Statistics and Society, *Walter T. Federer*
Vol. 4: Multivariate Analysis: A Selected and Abstracted Bibliography, 1957-1972, *Kocherlakota Subrahmaniam and Kathleen Subrahmaniam* (out of print)
Vol. 5: Design of Experiments: A Realistic Approach, *Virgil L. Anderson and Robert A. McLean*
Vol. 6: Statistical and Mathematical Aspects of Pollution Problems, *John W. Pratt*
Vol. 7: Introduction to Probability and Statistics (in two parts), Part I: Probability; Part II: Statistics, *Narayan C. Giri*
Vol. 8: Statistical Theory of the Analysis of Experimental Designs, *J. Ogawa*
Vol. 9: Statistical Techniques in Simulation (in two parts), *Jack P. C. Kleijnen*
Vol. 10: Data Quality Control and Editing, *Joseph I. Naus* (out of print)
Vol. 11: Cost of Living Index Numbers: Practice, Precision, and Theory, *Kali S. Banerjee*
Vol. 12: Weighing Designs: For Chemistry, Medicine, Economics, Operations Research, Statistics, *Kali S. Banerjee*
Vol. 13: The Search for Oil: Some Statistical Methods and Techniques, *edited by D. B. Owen*
Vol. 14: Sample Size Choice: Charts for Experiments with Linear Models, *Robert E. Odeh and Martin Fox*
Vol. 15: Statistical Methods for Engineers and Scientists, *Robert M. Bethea, Benjamin S. Duran, and Thomas L. Boullion*
Vol. 16: Statistical Quality Control Methods, *Irving W. Burr*
Vol. 17: On the History of Statistics and Probability, *edited by D. B. Owen*
Vol. 18: Econometrics, *Peter Schmidt*
Vol. 19: Sufficient Statistics: Selected Contributions, *Vasant S. Huzurbazar (edited by Anant M. Kshirsagar)*
Vol. 20: Handbook of Statistical Distributions, *Jagdish K. Patel, C. H. Kapadia, and D. B. Owen*
Vol. 21: Case Studies in Sample Design, *A. C. Rosander*
Vol. 22: Pocket Book of Statistical Tables, *compiled by R. E. Odeh, D. B. Owen, Z. W. Birnbaum, and L. Fisher*
Vol. 23: The Information in Contingency Tables, *D. V. Gokhale and Solomon Kullback*

Vol. 24: Statistical Analysis of Reliability and Life-Testing Models: Theory and Methods, *Lee J. Bain*
Vol. 25: Elementary Statistical Quality Control, *Irving W. Burr*
Vol. 26: An Introduction to Probability and Statistics Using BASIC, *Richard A. Groeneveld*
Vol. 27: Basic Applied Statistics, *B. L. Raktoe and J. J. Hubert*
Vol. 28: A Primer in Probability, *Kathleen Subrahmaniam*
Vol. 29: Random Processes: A First Look, *R. Syski*
Vol. 30: Regression Methods: A Tool for Data Analysis, *Rudolf J. Freund and Paul D. Minton*
Vol. 31: Randomization Tests, *Eugene S. Edgington*
Vol. 32: Tables for Normal Tolerance Limits, Sampling Plans, and Screening, *Robert E. Odeh and D. B. Owen*
Vol. 33: Statistical Computing, *William J. Kennedy, Jr. and James E. Gentle*
Vol. 34: Regression Analysis and Its Application: A Data-Oriented Approach, *Richard F. Gunst and Robert L. Mason*
Vol. 35: Scientific Strategies to Save Your Life, *I. D. J. Bross*
Vol. 36: Statistics in the Pharmaceutical Industry, *edited by C. Ralph Buncher and Jia-Yeong Tsay*
Vol. 37: Sampling from a Finite Population, *J. Hajek*
Vol. 38: Statistical Modeling Techniques, *S. S. Shapiro*
Vol. 39: Statistical Theory and Inference in Research, *T. A. Bancroft and C.-P. Han*
Vol. 40: Handbook of the Normal Distribution, *Jagdish K. Patel and Campbell B. Read*
Vol. 41: Recent Advances in Regression Methods, *Hrishikesh D. Vinod and Aman Ullah*
Vol. 42: Acceptance Sampling in Quality Control, *Edward G. Schilling*
Vol. 43: The Randomized Clinical Trial and Therapeutic Decisions, *edited by Niels Tygstrup, John M. Lachin, and Erik Juhl*
Vol. 44: Regression Analysis of Survival Data in Cancer Chemotherapy, *Walter H. Carter, Jr., Galen L. Wampler, and Donald M. Stablein*
Vol. 45: A Course in Linear Models, *Anant M. Kshirsagar*
Vol. 46: Clinical Trials: Issues and Approaches, *edited by Stanley H. Shapiro and Thomas H. Louis*
Vol. 47: Statistical Analysis of DNA Sequence Data, *edited by B. S. Weir*
Vol. 48: Nonlinear Regression Modeling: A Unified Practical Approach, *David A. Ratkowsky*
Vol. 49: Attribute Sampling Plans, Tables of Tests and Confidence Limits for Proportions, *Robert E. Odeh and D. B. Owen*
Vol. 50: Experimental Design, Statistical Models, and Genetic Statistics, *edited by Klaus Hinkelmann*
Vol. 51: Statistical Methods for Cancer Studies, *edited by Richard G. Cornell*
Vol. 52: Practical Statistical Sampling for Auditors, *Arthur J. Wilburn*
Vol. 53: Statistical Signal Processing, *edited by Edward J. Wegman and James G. Smith*
Vol. 54: Self-Organizing Methods in Modeling: GMDH Type Algorithms, *edited by Stanley J. Farlow*
Vol. 55: Applied Factorial and Fractional Designs, *Robert A. McLean and Virgil L. Anderson*
Vol. 56: Design of Experiments: Ranking and Selection, *edited by Thomas J. Santner and Ajit C. Tamhane*
Vol. 57: Statistical Methods for Engineers and Scientists. Second Edition, Revised and Expanded, *Robert M. Bethea, Benjamin S. Duran, and Thomas L. Boullion*
Vol. 58: Ensemble Modeling: Inference from Small-Scale Properties to Large-Scale Systems, *Alan E. Gelfand and Crayton C. Walker*
Vol. 59: Computer Modeling for Business and Industry, *Bruce L. Bowerman and Richard T. O'Connell*
Vol. 60: Bayesian Analysis of Linear Models, *Lyle D. Broemeling*

Vol. 61: Methodological Issues for Health Care Surveys, *Brenda Cox and Steven Cohen*

Vol. 62: Applied Regression Analysis and Experimental Design, *Richard J. Brook and Gregory C. Arnold*

Vol. 63: Statpal: A Statistical Package for Microcomputers – PC-DOS Version for the IBM PC and Compatibles, *Bruce J. Chalmer and David G. Whitmore*

Vol. 64: Statpal: A Statistical Package for Microcomputers – Apple Version for the II, II+, and IIe, *David G. Whitmore and Bruce J. Chalmer*

Vol. 65: Nonparametric Statistical Inference, Second Edition, Revised and Expanded, *Jean Dickinson Gibbons*

Vol. 66: Design and Analysis of Experiments, *Roger G. Petersen*

Vol. 67: Statistical Methods for Pharmaceutical Research Planning, *Sten W. Bergman and John C. Gittins*

Vol. 68: Goodness-of-Fit Techniques, *edited by Ralph B. D'Agostino and Michael A. Stephens*

Vol. 69: Statistical Methods in Discrimination Litigation, *edited by D. H. Kaye and Mikel Aickin*

Vol. 70: Truncated and Censored Samples from Normal Populations, *Helmut Schneider*

Vol. 71: Robust Inference, *M. L. Tiku, W. Y. Tan, and N. Balakrishnan*

Vol. 72: Statistical Image Processing and Graphics, *edited by Edward J. Wegman and Douglas J. DePriest*

Vol. 73: Assignment Methods in Combinatorial Data Analysis, *Lawrence J. Hubert*

Vol. 74: Econometrics and Structural Change, *Lyle D. Broemeling and Hiroki Tsurumi*

Vol. 75: Multivariate Interpretation of Clinical Laboratory Data, *Adelin Albert and Eugene K. Harris*

Vol. 76: Statistical Tools for Simulation Practitioners, *Jack P. C. Kleijnen*

Vol. 77: Randomization Tests, Second Edition, *Eugene S. Edgington*

Vol. 78: A Folio of Distributions: A Collection of Theoretical Quantile-Quantile Plots, *Edward B. Fowlkes*

Vol. 79: Applied Categorical Data Analysis, *Daniel H. Freeman, Jr.*

Vol. 80: Seemingly Unrelated Regression Equations Models : Estimation and Inference, *Virendra K. Srivastava and David E. A. Giles*

Vol. 81: Response Surfaces: Designs and Analyses, *Andre I. Khuri and John A. Cornell*

Vol. 82: Nonlinear Parameter Estimation: An Integrated System in BASIC, *John C. Nash and Mary Walker-Smith*

Vol. 83: Cancer Modeling, *edited by James R. Thompson and Barry W. Brown*

Vol. 84: Mixture Models: Inference and Applications to Clustering, *Geoffrey J. McLachlan and Kaye E. Basford*

Vol. 85: Randomized Response: Theory and Techniques, *Arijit Chaudhuri and Rahul Mukerjee*

Vol. 86: Biopharmaceutical Statistics for Drug Development, *edited by Karl E. Peace*

Vol. 87: Parts per Million Values for Estimating Quality Levels, *Robert E. Odeh and D. B. Owen*

Vol. 88: Lognormal Distributions: Theory and Applications, *edited by Edwin L. Crow and Kunio Shimizu*

Vol. 89: Properties of Estimators for the Gamma Distribution, *K. O. Bowman and L. R. Shenton*

Vol. 90: Spline Smoothing and Nonparametric Regression, *Randall L. Eubank*

Vol. 91: Linear Least Squares Computations, *R. W. Farebrother*

Vol. 92: Exploring Statistics, *Damaraju Raghavarao*

Vol. 93: Applied Time Series Analysis for Business and Economic Forecasting, *Sufi M. Nazem*

Vol. 94: Bayesian Analysis of Time Series and Dynamic Models, *edited by James C. Spall*

Vol. 95: The Inverse Gaussian Distribution: Theory, Methodology, and Applications, *Raj S. Chhikara and J. Leroy Folks*

Vol. 96: Parameter Estimation in Reliability and Life Span Models, *A. Clifford Cohen and Betty Jones Whitten*

Vol. 97: Pooled Cross-Sectional and Time Series Data Analysis, *Terry E. Dielman*

Vol. 98: Random Processes: A First Look, Second Edition, Revised and Expanded, *R. Syski*

Vol. 99: Generalized Poisson Distributions: Properties and Applications, *P.C. Consul*

Vol. 100: Nonlinear L_p-Norm Estimation, *René Gonin and Arthur H. Money*

Vol. 101: Model Discrimination for Nonlinear Regression Models, *Dale S. Borowiak*

Vol. 102: Applied Regression Analysis in Econometrics, *Howard E. Doran*

Vol. 103: Continued Fractions in Statistical Applications, *K.O. Bowman and L.R. Shenton*

Vol. 104: Statistical Methodology in the Pharmaceutical Sciences, *Donald A. Berry*

Vol. 105: Experimental Design in Biotechnology, *Perry D. Haaland*

Vol. 106: Statistical Issues in Drug Research and Development, *edited by Karl E. Peace*

Vol. 107: Handbook of Nonlinear Regression Models, *David A. Ratkowsky*

Vol. 108: Robust Regression: Analysis and Applications, *edited by Kenneth D. Lawrence and Jeffrey L. Arthur*

Vol. 109: Statistical Design and Analysis of Industrial Experiments, *edited by Subir Ghosh*

Vol. 110: *U*-Statistics: Theory and Practice, *A. J. Lee*

Vol. 111: A Primer in Probability, Second Edition, Revised and Expanded, *Kathleen Subrahmaniam*

Vol. 112: Data Quality Control: Theory and Pragmatics, *edited by Gunar E. Liepins and V. R. R. Uppuluri*

Vol. 113: Engineering Quality by Design: Interpreting the Taguchi Approach, *Thomas B. Barker*

Vol. 114: Survivorship Analysis for Clinical Studies, *Eugene K. Harris and Adelin Albert*

ADDITIONAL VOLUMES IN PREPARATION

DATA QUALITY CONTROL

Theory and Pragmatics

Edited by

GUNAR E. LIEPINS
V. R. R. UPPULURI

Oak Ridge National Laboratory
Oak Ridge, Tennessee

MARCEL DEKKER, INC. New York • Basel • Hong Kong

Library of Congress Cataloging-in-Publication Data

Data quality control : theory and pragmatics / edited by Gunar Liepins, V. R. R. Uppuluri.
p. cm. — (Statistics, textbooks and monographs ; vol. 112)
"This book has its roots in the findings of the Survey Data Quality Control Workshop which was held in Oak Ridge, Tennessee, on April 21–22, 1988. The workshop was sponsored jointly by Oak Ridge National Laboratory, the American Statistical Association, and the Data Systems Research and Development Program of Martin Marietta Energy Systems, Inc."—Pref.
Includes bibliographical references.
ISBN 0-8247-8354-9 (alk. paper)
1. Mathematical statistics—Data processing—Quality control. I. Liepins, G. E. II. Uppuluri, V. R. R. III. Survey Data Quality Control Workshop (1988 : Oak Ridge, Tenn.) IV. Oak Ridge National Laboratory. V. American Statistical Association. VI. Martin Marietta Energy Systems, Inc. Data Systems Research and Development Program. VII. Series: Statistics, textbooks and monographs ; v. 112.
QA276.4.D36 1990
519.5′0285—dc20 90-34507
CIP

This book is printed on acid-free paper.

MARCEL DEKKER, INC.
270 Madison Avenue, New York, New York 10016

Current printing (last digit):
10 9 8 7 6 5 4 3 2 1

PRINTED IN THE UNITED STATES OF AMERICA

Preface

This book is based on two important premises: First, many useful data quality control practices have gone undocumented; publication of these practices would be a service to the quality control community and could help improve data quality. Second, better interaction among quality control practitioners, managers, and theoreticians would lead to a profitable exchange of ideas. Theoreticians and managers would better understand the real concerns of the practitioners. Practitioners and managers would better understand the potential applications and implications of the theory being developed. Theoreticians and practitioners would better understand the priorities held by managers. This would enable each group to understand the role of the other groups in working toward the mutual goal of improved data quality.

A workshop was held in April, 1988, in Oak Ridge, Tennessee, to foster communication and interchange among the three groups and to identify central data quality issues and recent solutions. A straw man data quality control cycle was circulated among the attendees prior to the workshop to serve as an initial focal point. After the workshop, attendees and selected additional authors were requested to coordinate their thoughts about research, practice, and management of data quality.

What evolved from the thoughtful contributions of the authors has much broader implications than the original straw man data quality control cycle. Although not all of the topics received equal attention, each of the following topics is addressed in this book: Which current data quality control practices are successful? Is the current level of data adequate? What current research results might be applied to further improve data quality? What are the barriers to improved data quality? How can any such barriers be removed? What additional research needs to be pursued? What management steps could be taken to promote data quality improvement? What can practitioners and theoreticians do to provide the concerned manager with the support he or she needs?

The book is organized around these topics in 16 chapters. The first two chapters provide a history of data quality control issues as perceived by a senior manager and a senior government official, respectively. Chapter 1 succinctly summarizes important findings of over a decade of data validation efforts on behalf of the federal government. In addition, the very important point is made that data quality is more than just a technical matter; practical and organizational issues need to be addressed. Chapter 2 reviews the progress in data quality control since the study by the Panel on Incomplete Data. It suggests that the lag between research and application in data quality can all too frequently be attributed to lack of a clear economic incentive for quality data. It further identifies the characteristics and strategies currently required of successful proponents of data quality.

The third chapter briefly reviews some of the findings from one of the validation studies undertaken by Oak Ridge National Laboratory (ORNL) for the Energy Information Administration (EIA). It specifies the correspondence between the stages of the study and the steps of the proposed quality control cycle.

Chapters 4 through 8 detail actual quality control practices that have been and continue to be successful in industry as well as government. Chapter 4 describes successful quality control procedures implemented at United States Fidelity and Guaranty Company. Three types of audits are described. Systems audits determine whether systems are operating as designed, statistical audits determine and monitor data accuracy, and procedural audits determine whether all parts of the company are performing quality control in accordance with programmatic guidelines. Chapter 5 describes quality control procedures for shared multidisciplinary databases. Particularly noteworthy is the recognized need to provide incentives to each of the multiple user groups to insure maximum possible data quality. The incentives used are detailed. Equally important is the database design approach that explicitly includes metadata. Chapter 6 suggests data quality control practices that are useful in the collection and maintenance of a data stream over multiple decades. Sampling techniques receive special emphasis. Chapter 7 discusses data quality requirements for and their impacts on automobile warranty decisions. Because collecting the required data directly is difficult, the need for corroborating data is great. Such corroborating data are collected from field sources, including proving–ground tests, surveys, and direct warranty data. How the data are reconciled and limitations to the approach are discussed. Chapter 8 describes quality assurance steps taken by Westat Inc., a company that designs and implements surveys. Sample design and selection, instrument design, quality in field operations, recruiting and training, sampling, interviewing, corrective actions, data preparation, receipt control and field audit, keying, machine editing, and standardization are all addressed.

Chapter 9 provides a transition from practice to theory. It describes a "field theory" of data, summarizes the validation studies performed by ORNL for the EIA, and concludes that for federally collected energy data, data quality problems arise principally either when there is an incentive to private companies to provide misleading data, or when valid data are used for purposes for which they were not intended.

The remaining seven chapters are technical and address editing, imputation, statistical matching, and error localization. Chapter 10 provides an overview of techniques to minimize post-collection adjustments to data, as well as a survey of editing and imputation techniques when such adjustments are required. Main topics include design of data collection efforts to limit impact of nonsampling errors, the role of statistical models in refining editing and nonresponse adjustments, univariate and multivariate models for outlier detection, data analysis in the presence of outliers, and recent modeling literature. In spite of the attention devoted to technical adjustments, data modification is clearly stated to be a last resort. It is far better to have quality original data than to rely on technical adjustments.

Chapter 11 provides recommendations for collection of quality data about rare or sensitive items. Network sampling is defined, described, and recommended. The chapter illustrates the usefulness of network sampling with an example of its application to a survey about missing children.

Chapter 12 discusses the use of supplementary variables to obtain unbiased estimates of statistics of interest about primary variables in the presence of nonresponse. Chapter 13 describes software for the analysis of survey data.

Chapter 14 illustrates the use of Bayesian statistics to make inferences about the proportion of defective items in a population. Although the example given is an application to manufacturing, the technique should be extendable to inferences about the data quality. The authors state that an advantage of the Bayesian technique over traditional techniques is its computational directness.

Chapter 15 provides a detailed account of how to merge data sets when the data of interest are contained in no single data set, but could possibly be reconstructed from a variety of different data sets. (As repeatedly stated in the previous chapters, artificial construction of data is a last resort to be used, for example, when direct collection of data is either economically or temporally infeasible.) The results of recent comparative experiments with two versions of unconstrained matching versus optimally constrained matching, different distance functions (used to determine matches), and varying amounts of noise and bias are reported.

Chapter 16 provides an analysis of the effectiveness of one error localization technique, Minimum Weighted Fields to Impute (MWFI). Results reinforce the conclusion that technical fixes cannot overcome inherent weaknesses in the data; the worse the data, the less useful the fixes.

To conclude the preface here would leave the reader guessing, What exactly was the straw man data quality control cycle and how was it received at the workshop? This preface concludes with the answer to that question.

The data quality control cycle for periodically collected data is defined in seven steps:

(a) Review, design and pretest
(b) Collect
(c) Screen and edit
(d) Evaluate economics
(e) Localize error
(f) Analyze
(g) Impute and reinterview

Three observations are noteworthy:

1. Incorporated into the cycle are the three traditional analytical data editing steps: edit (check for consistency), localize error (infer which fields are wrong in inconsistent records), and impute (inject values into the missing or incorrect fields to render each record complete and consistent).

2. The subset of the cycle demarked by dotted lines would seem to

be appropriate for development as an expert system, a system that would query the practitioner about his data characteristics and the expected use and would suggest appropriate techniques and procedures to be followed.

3. Ideally, the whole data quality control cycle would be incrementally applied: diagnosis of problems at an early stage of the data stream may help alleviate later problems.

Presentations at the workshop addressed most of the topics suggested by the quality control cycle. Findings could be roughly characterized as either technical or institutional/motivational. The degree to which the quality of a data system is dependent on the solution of institutional issues was suggested to be inversely related to the degree of control exercised by statisticians over the system. At one end of the spectrum, the quality of the data of the Framingham Heart Study seemed to be determined primarily by technical factors (Anderson, Dumphy and Wilson, Chapter 6). At the other end of the spectrum, quality of the Department of Energy's data requires that complex organizational and motivational issues be solved (Hale, Chapter 2). Consensus seemed to be reached on four technical points:

1. Simple quality control measures should be pursued before more difficult and costly steps are taken.
2. Data errors and biases should be caught at the earliest possible stage of the quality control cycle (Little, Chapter 10).
3. Software should be standardized (LaVange and Shah, Chapter 13).
4. The development of appropriate (understandable and useful) summary measures and guidelines for data usage remains a central open research issue (Loebl, Chapter 9).

The strongest criticism of the data quality control cycle was directed at its lack of explicit attention to organizational issues. Large, federally administered data systems require the cooperation of many individuals from a diversity of organizational units. How can individuals in the data collection, processing, and publication activities be motivated to be personally accountable for the quality of the data? How can managers be convinced of the usefulness of statistics and statisticians? How can statisticians be convinced of the importance of thoroughly understanding the system and its requirements as opposed to pursuing statistics for statistics' sake? How can statistical training and professional recognition be redirected to better prepare and motivate statisticians for data quality control? How can incentives be structured to promote total quality

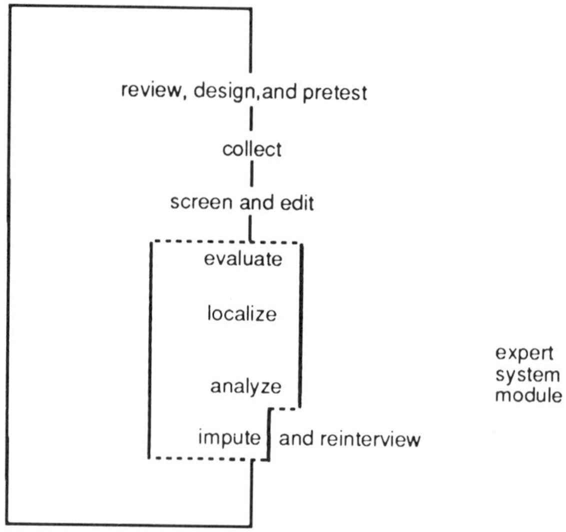

Data Quality Cycle

control at all organizational levels? Flournoy and Hearns (Chapter 5) provide some answers. Perhaps not surprisingly, the major conclusion of the workshop seemed to be that data system managers need to be highly perseverant, imaginative, forceful, and persuasive in motivating and educating about the benefits of total quality control. To be most effective, they should structure data systems so that all parties to the systems (respondents, data compilers, analysts and statisticians, and data users) share a vested interest in its quality. Within large organizations this is no mean task. How to promote awareness of the need for total data quality control, foster the required organizational changes, and train and motivate data quality control professionals are long-term issues that the statistical community must continue to address. We hope that this book will increase the awareness of these issues and provide one step towards their resolution.

Gunar E. Liepins
V. R. R. Uppuluri

Contents

Contributors

KEAVEN M. ANDERSON Framingham Heart Study, National Heart, Lung, and Blood Institute, Framingham, Massachusetts

BETH G. BARNWELL Research Triangle Institute, Research Triangle Park, North Carolina

RICHARD S. BARR Southern Methodist University, Dallas, Texas

JOHNNY BLAIR* Survey Research Laboratory, University of Illinois at Urbana-Champaign, Urbana, Illinois

DEBORAH C. DUMPHY Framingham Heart Study, National Heart, Lung, and Blood Institute, Framingham, Massachusetts

NANCY FLOURNOY The American University, Washington, DC.

DOUGLAS R. HALE Energy Information Administration, Department of Energy, Washington, D.C.

MORRIS H. HANSEN Westat, Inc., Rockville, Maryland

LEONARD B. HEARNE George Mason University, Fairfax, Virginia

JOYCE F. KILLINGER Research Triangle Institute, Research Triangle Park, North Carolina

**Current affiliation*: Survey Research Center, University of Maryland College Park, College Park, Maryland.

LISA M. LaVANGE* Research Triangle Institute, Research Triangle Park, North Carolina

NANCY J. LePAGE United States Fidelity and Guaranty Company, Baltimore, Maryland

GUNAR E. LIEPINS Energy Division, Oak Ridge National Laboratory, Oak Ridge, Tennessee

RODERICK J. LITTLE University of California at Los Angeles, Los Angeles, California

ANDREW S. LOEBL Martin Marietta Energy Systems, Inc., Oak Ridge, Tennessee

GARY C. McDONALD General Motors Research Laboratories, Warren, Michigan

DAVID R. MORGANSTEIN Westat, Inc., Rockville, Maryland

D. J. PACK School of Business Administration, Miami University, Oxford, Ohio

T. E. RAGHUNATHAN University of Washington, Seattle, Washington

PODURI S. R. S. RAO University of Rochester, Rochester, New York

JEFFREY A. ROBINSON General Motors Research Laboratories, Warren, Michigan

DONALD B. RUBIN Harvard University, Cambridge, Massachusetts

BABUBHAI V. SHAH Research Triangle Institute, Research Triangle Park, North Carolina

J. SCOTT TURNER College of Business Administration, Oklahoma State University, Stillwater, Oklahoma

PETER W. F. WILSON Framingham Heart Study, National Heart, Lung, and Blood Institute, Framingham, Massachusetts

*_Current affiliation_: Frank Porter Graham Child Development Center, University of North Carolina, Chapel Hill, North Carolina.

DATA QUALITY CONTROL

1

An Organizational and Historical Perspective of a Decade of Data Validation R&D at the Oak Ridge Reservation

ANDREW S. LOEBL Martin Marietta Energy Systems, Inc., Oak Ridge, Tennessee

As head of the Data Systems Research and Development (DSRD) Program of Martin Marietta Energy Systems, Inc., I was approached by the chairs to cosponsor this Survey Data Quality Control workshop. Although DSRD is a relatively young organization (eight years), it sprang from a variety of earlier data quality/data validation efforts and I judged it entirely appropriate to support the workshop. I felt that it was important to publicly summarize and document the lessons learned over the past decade. As you will note from the program, the other cosponsor of the workshop is Oak Ridge National Laboratory (ORNL). ORNL's objectives will be discussed in a later presentation, so I will limit my remarks to a summary of DSRD program issues and lessons learned, especially from an organizational standpoint.

To help focus this conference, and provoke response, a proposed quality control cycle was outlined and described in the workshop announcement. I do not intend to comment directly on the appropriateness of the cycle; I am not a technical person. Whether the concept of the quality control cycle is technically sound and whether its implementation is feasible are matters that you, the workshop conferees, will debate. Rather, I will discuss certain issues from the perspective of my belief that quality assurance of data goes beyond classical approaches which measure data accuracy in terms of selection error, respondent error, processing error, and, most particularly, specification error. My personal experience spans data problems associated with a broad range of programs (statistical and analytical): census, energy, logistics, personnel, fiscal, health, intelligence, environment, defense operations, labor, material, research, command and control, engineering, security, business management, etc. My remarks are the result of reflecting on and coalescing from an experience with data problems of over 70 agencies and departments of the federal government since 1975.

Perhaps, in view of my disclaimers, it is surprising that my comments ultimately reflect back on the data quality control cycle. Regardless of its technical adequacy, the cycle does not explicitly address the organizational issues of data quality, and the organizational issues are at least as complex and important as the technical issues. To allow you to see what has led me to this conclusion, let me sketch a brief history of my organization's involvement with data quality.

My organization began its involvement with data quality just after the energy shocks of 1973 and 1974. At that time, it was deemed important to know and understand energy use patterns across all sectors of the United States economic structure. In response to this need, three compendiums of energy use data were begun: *The Transportation Energy Use Data Book*, *The Industrial Energy Use Data Book*, and the *Buildings Energy Use Data Book*. An example of the type of data presented in the last volume is the inventory of buildings by type, economic sector, geographical location, and primary space heating energy source across the United States. As I saw our charter, our responsibilities were to:

Evaluate the need for and importance of data items
Locate and compile data from public and private sources
Analyze and annotate the data

We were not charged with collecting data from the original respondents, nor with informing legislative bodies regarding the availability and validity of classes of data. We compiled, interpreted, and explained important available data. In short, we played something of a middleman role between raw data collection and the end user (see Figure 1).

Insofar as we did not interact with those organizations that mandated data collection, nor with their respondents, our appreciation of programmatic and organizational issues was simplistic. We had our own constituency and diverse sources of disparate data. Our major finding was that all too frequently the data trail was informal. It was difficult to trace

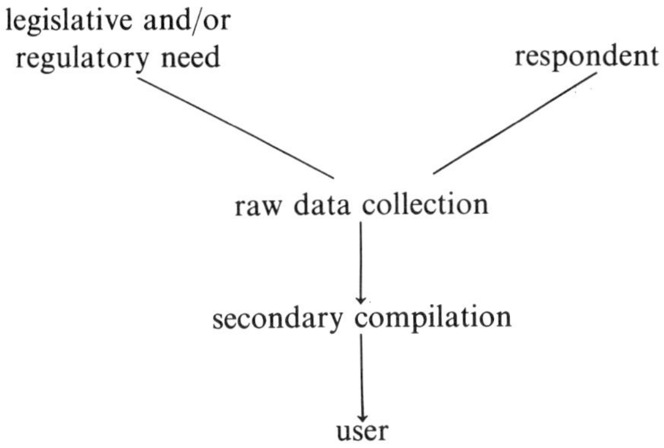

Figure 1. The flow of communication in data collection and use as experienced in DSRD while compiling energy use data books.

transformations that the data had undergone and document the caveats that were applicable. Individuals at the data collection agencies seemed industrious and conscientious, but the institutional memory seemed all too frequently to reside in a single individual or small group of experts. A systematic policy of quality control was often missing.

With the inception of the Energy Information Administration, the role of my organization changed. No longer were we charged with secondary compilation of data; we were now responsible for data validation. That is, we were responsible for the following tasks:

Requirements review
Identification of users
Auditing accuracy and appropriateness
Verifying frames
Documenting respondent burden
Evaluating, editing, and imputation methods
Determining similarity with other data collection efforts
Reducing redundancy and the volume of data
Reconciling various conflicting data series

Oak Ridge brought two significant perspectives to the tasks. First, our technical and institutional independence brought a measure of objectivity to what theretofore had been a frenetic recitation of symptoms. The laboratory's reputation for deliberate, sober reflection quelled enough of the furor so that work could begin on dealing with the issues, rather than merely decrying them. Second, the laboratory brought to the problem a mix of administrative and scientific skill which merged data users' concerns for data quality with program managers' concerns for performance. Our dual strengths in research (as data users) and research management (as program supervisors) enabled the data validation effort to address the concerns of (1) those who lamented the quality of energy data, and (2) those who were responsible for improving the quality of energy data. The impetus of our program continues to focus, as always, upon total data quality.

With this new charter (in 1977), we no longer found ourselves removed from the respondents and the agencies responsible for the raw data collection; we dealt intimately with them (see Figure 2). Equally important, we found ourselves supporting the users of data. As a result, our perspective became less simplistic. Nonetheless, our fundamental findings remained much the same.

After many tens of thousands of person-hours devoted to investigating and validating a variety of federal data collection efforts and their associated systems, we were able to establish that although accuracy varied according to data type and data collection instrument, most data were reasonably accurate (to the best of our ability to determine). Unfortunately, a few findings were less reassuring. We discovered that for certain of the data, we could locate no direct users. Similarly, although the data were often accurate, at times they were not the data that actually were needed.

Much the same findings resurfaced year after year as we completed studies of nonenergy data and models for both civilian and noncivilian agencies and authorities. We found a data base that had great potential

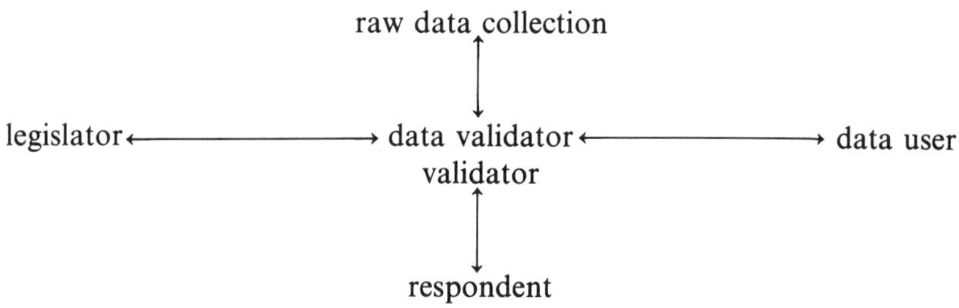

Figure 2. Data validation interacts with agencies, authorities, and users at all stages of data collection and use.

use, but whose architecture seemed to have been chosen for historical reasons that no one clearly recalled. The data base certainly was a repository for data, but unfortunately under these circumstances, the user could neither retrieve from it the information that he needed, retain confidence in the resource, nor reconcile his need for the data with the poorly understood mandate for its original collection.

This seems to raise a large issue for quality control: the need for communication and feedback among the different organizations that interact with the data. Each organization needs to pursue a realistic, systematic practice of quality control (in very broad and not purely statistical or empirical ways).

So far, I can report some important results of our work which seem to be relevant to some of the more pragmatic data quality issues we have encountered. The first set of results can be reported as three major conclusions:

1. While every data collection system is potentially subject to random errors, the systematic errors in large-scale data collection systems are much more significant than the random errors.
2. While bias or misrepresentation are inevitably possible, they are likely to cause massive distortion only when respondents consciously report misleading data for the purpose of gaining some kind of advantage.
3. The single most significant source of error in data analysis is misapplication of data which would be reasonably accurate in the right context.

The second category of results can be characterized as a general field theory of data. The argument I make is that data ought to be reviewed as belonging to a field wherein they can have both accuracy and (changing) relevance characteristics. I believe that since both characteristics are essential for a valid analysis, and since accuracy is the customary preoccupation of analysts, it follows that (lack of) relevance is the most likely source of data difficulties.

At the federal level, the quality control objectives (of interorganizational awareness of the need for systematic total data quality control, beginning with the mandate for data collection and continuing through the data analysis) seem virtually unattainable. The federal government spends between $17 billion and $50 billion to collect, organize, maintain, and properly employ data. Still, the critical, purposeful interaction between the various agencies and individuals responsible for mandating the data collection, collecting the data, and reviewing, publishing, and using the data seems often to be lost in internal rivalries, jurisdictional indetermination, lack of immediate economic incentive, and the overwhelming enormity of the data quality control problem.

One can justifiably say that we are not embarking on the information era, we are in the data age. Perhaps there are insights that you, the attendees of this workshop, could share to help bring us one step closer to the critical data quality objectives we need to reach. I leave you with this challenge: While discussing the technical issues of survey data quality control, do not leave the practical and organizational issues ignored.

I offer the following thoughts in closing:

"Data Validation is the process of determining how strong and authoritative data is for its intended uses."—Andrew Loebl

"Nothing is true or false, just more or less useful."—David Easton

2
From Research to Application

DOUGLAS R. HALE Energy Information Administration, Department of Energy, Washington, D.C.

1. INTRODUCTION

This conference on data quality control is something of a celebration for me. Nine years ago, many of us met at the Energy Information Administration (EIA) in an effort to validate energy data and forecasts. Andy Loebl and Gordon Sande, and a few others not here today, were the first to help me in that task. The issue then was data integrity, not data accuracy. Energy production, price, and consumption numbers were going to the Council of Economic Advisors and to the White House and no one knew their origins. We do not talk about data integrity any more because hard work—much of it done by Oak Ridge National Laboratory (ORNL) researchers—and persistence—mostly on the part of three EIA administrators and my boss, Yvonne M. Bishop—paid off.

I do not want to give you the impression that the journey from data integrity to data quality was either inevitable or even pleasant. Until this morning, the last time I saw Andy was four years ago in the Deputy Administrator's office. We were screaming at each other over who botched some project or other. We have come far, and I am enormously proud of that. This workshop is a good place to think about the next steps in improving data quality.

The purposes of this workshop are:

1. To summarize the progress in survey data quality control since the Panel on Incomplete Data's study*
2. To evaluate the concept of a survey data quality control cycle
3. To identify future research needs and applications, in civilian statistical agencies.

My main points are:

1. Application of the panel's good advice has been unnecessarily slow.
2. To be successful, the quality control cycle must explicitly cope with the newer obstacles to research and application in civilian statistical agencies.
3. A shortage of worthwhile research ideas and high-payoff applications is not the problem. The problem is to realize their potential.

My remarks are from the point of view of a senior government official charged with improving the quality of energy data and forecasts. But what I have to say today is strictly personal—it does not reflect the policy of the Department of Energy (DOE). The chairman asked for an exchange of views, not for policy positions, and that is what he will get.

2. PANEL ON INCOMPLETE DATA

The Panel on Incomplete Data's three-volume report was published in 1983. As its name implies, the panel focused on incomplete data as a source of survey error (Table 1). It did not stress sampling, response, or processing error. Within the area of incomplete data, the panel carefully reviewed theoretical and practical ways to deal with unit and item nonresponse. The panel did not address undercoverage.

As shown in Table 2, the panel made 24 recommendations for good statistical practice. The first recommendation is to collect accurate, complete data in the first place. The remaining 23 recommendations suggest what to do when that fails and address data forms, database design, measurement and management of nonresponse, analysis, and documentation. I have grouped the recommendations into six categories to make the ensuing discussion more manageable. The crosswalk from the

*Panel on Incomplete Data, *Incomplete Data in Sample Surveys*, Academic Press (ISBN 0-12-363901-8), New York, 1983.

Table 1. Sources of Error

Sampling	
	Undercoverage
Incomplete data	Unit nonresponse
	Item nonresponse
Response	
Processing	

panel's recommendations to these categories is contained in the appendix to this chapter.

Five years later, the Office of Management and Budget (OMB) issued draft guidelines for statistical agencies*. When final, these guidelines, which have been in development for over two years, will establish the minimal requirements that federal data collections must meet. OMB can disallow any data collection that fails these requirements.

Table 2. Good Statistical Practice

Panel Recommendations (1983)	OMB Draft Guidelines (1988)
Get the data (1)	x
Good collection, form, and database design (6)	x
Measure and manage nonresponse (4)	x
Design survey for nonresponse (2)	x
Analyze (9)	
Effect of imputation (4)	x
Alternative schemes (5)	
Documentation (2)	
Survey	x
Effect of incomplete data	x

*Office of Management and Budget, Guidelines for Federal Statistical Activities; Notice, *Part II*, *Federal Register*, January 20, 1988, pp. 1542–1552.

It is a compliment to the panel that OMB has adopted essentially all its suggestions. Being minimal standards, OMB has not asked for all the information that the panel suggested. For instance, Table 2 shows that OMB does not require agencies to show that they have evaluated alternative imputation schemes. OMB requires only that the scheme be identified and its impact on the data shown. In all other respects, OMB has adopted the spirit of the panel's ideas.

The panel also made 11 research suggestions, which I have grouped into six categories in Table 3: software standardization, nonresponse investigation, imputation, terminology, cost effectiveness, and effect on subsequent analysis. The crosswalk is contained in the appendix to this chapter.

With this background, I looked at how well my agency, the EIA, is doing in carrying out the panel's recommendations. EIA is a small, civilian agency of about 475 people that conducts about 70 surveys ranging from weekly to triennial in frequency. Most of the surveys are intended to be complete enumerations of energy firms, but EIA also conducts one population survey and one commercial building survey. To a greater or lesser degree, EIA faces all the incomplete data problems studied by the panel.

I was reassured to find that EIA is having difficulty carrying out the panel's suggestions in only three areas:

1. Nonresponse management
2. Analyzing the effects of incomplete data
3. Documenting the effects of incomplete data

The nonresponse-management category may be mysterious. EIA monitors unit nonresponse and item nonresponse primarily to guide its

Table 3. Panel's Research Recommendations (11)

Standardize software modules (1)
Investigate nonresponse mechanisms (4)
Evaluate imputation schemes (3)
Develop consistent terminology (1)
Determine cost-effectiveness of nonresponse procedures (1)
Investigate effect of incomplete data on complex analysis (1)

Table 4. Five Years Later

EIA Problem Area	Research Need
Nonresponse management	Software Cost-effectiveness data
Analyze incomplete data	
Effects	Software
Alternatives	Nonresponse mechanisms
Document effects	Standard terminology

collection efforts. These efforts are generally so successful that our problems are described more accurately as late response than as nonresponse. EIA often does not maintain response-rate series, and it rarely has individual response and imputation histories. Therefore, the data for analyzing patterns of nonresponse or for deciding the best imputation approach are not readily available. For similar reasons, the agency has difficulty reporting the extent of imputation in its more current numbers.

Having suffered through several expensive attempts to improve nonresponse analysis and evaluate imputation approaches, I easily identified my agency's perception of the research need underlying the problem. Table 4 relates the data quality problem area to the research need. Each of these research needs had been carefully described by the panel five years ago!

Five years is a long time, and there is no evidence that EIA is making any real progress in solving these problems. If substantial research progress has been made, then we either do not know about it or do not understand it.

3. RESEARCH AND APPLICATION LAGS

One response to these observations is to say that there have always been lags in going from research to application. Five years or ten years does not necessarily indicate a problem. That is true, but there is considerable evidence from the physical sciences that application lags are influenced by economics and can be controlled.

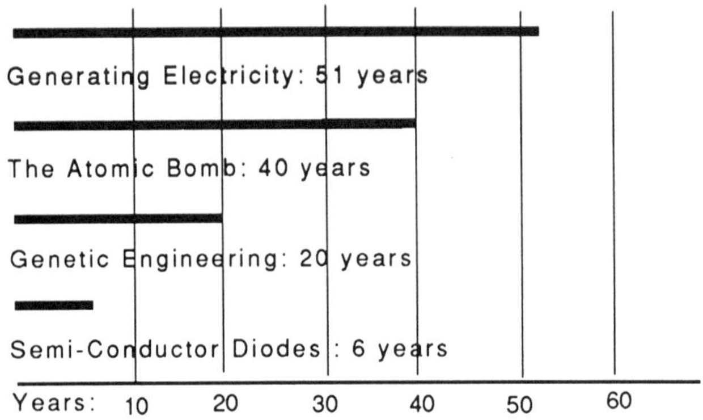

Figure 1. From basic science to practical use: the distance shrinks

Figure 1 is from a *New York Times* article discussing the length of time from the theoretical concept to the practical application of four discoveries.* The article supports the strong impression given in Figure 1 that the lags are getting much shorter. Recent patent research by Dr. Francis Narin of Computer Horizons, Inc., for the National Science Foundation also shows that the links to basic research, as measured by citations, became stronger between 1975 and 1986. All this seems to be a direct result of the maturing of traditional markets and increased international competition.

If we could somehow mimic economic competition in data quality research and application strategies, our problems would be reduced. I am not optimistic about that, but we can do much better by facing the obstacles and managing around them.

Table 5 shows my view of the major forces for and against data quality research and application in civilian statistical agencies. Without question, the greatest pressures pushing government managers to consider investing in research are the certainty of continued tight budgets and OMB's insistence on contracting all but "uniquely government" work. Good managers have applied all the proven remedies; new ideas are the only bet left. Contracting for survey operations is a subtle force because managers realize they cannot tell the quality or efficiency of work simply by looking at the product and the bill. To have any idea of the quality of

*Broad, William J., Science and Technology: The Gap is Shrinking Fast, *New York Times*, April 5, 1988, p. Cl.

Table 5. Data Quality

For:	Against:
Tight budgets	Economics of information
Contracting out	U.S. Treasury twist
Statistical guidelines	Limited enforcement
Court cases	Exceptional
Professionalism	Civil service
	Research culture

the data, good managers realize they must have extensive survey operations statistics, including those recommended by the panel.

Effectively countering these pressures for data quality research are the economics of information, augmented by a twist from the U.S. Treasury. Any manager investing in research is likely to be left with nothing but a bill. That is the nature of research. If the project pays off, who wins? The sponsor, to some extent—but the bigger winner is the manager down the hall who risked nothing and paid nothing, but who will gladly take what was found, maybe jazz it up slightly, and look like a hero. This "Japanese strategy" is brutally effective, but kills much research. The Treasury twist is this: Even if a manager saves money, it is taken away to fund someone else's folly—if he or she actually makes money, it goes into the public coffer.

One additional point related to the lack of advanced survey statistics is that developing information on data quality is expensive, and the final tables look the same whether the agency has quality information or not. There are clear incentives to cut corners instead of investing in better procedures.

Two positive forces for better data and research are OMB's statistical guidelines and court cases, however painful they may be. Unfortunately, these shocks are just too few and far between to change managers' perceptions.

Many managers still invest in data quality because they are professionals trying to do the right thing. All of us are sympathetic to those feelings. But the civil service and the research culture are not giving much comfort, as suggested below.

The ills of the civil service have undermined the traditional approaches to data quality research and application. In the old days, government analysts knew the literature, occasionally contributed to it, and quickly picked up the telephone to laboratory and university researchers. Supervisors and managers often had substantial technical knowledge. The government analysts were the eyes and ears; the laboratories and the analysts' old universities were the talent stables. In a bumbling sort of way, problems were recognized and measured against the standard of current knowledge, and research efforts were undertaken as needed.

Today, those government analysts are mostly gone. As for managers, the personnel system plainly views a senior manager with technical skills and interests as suspect, if not derelict. The EIA has not hired a Ph.D. mathematical statistician right out of school in three years. I am the only one to have hired Ph.D. mathematical economists into the agency within four years. Those analysts who are left have been forced by the pay and classification system to become managers—a disaster for both analysis and management, in my opimion.

All this would not necessarily be bad if researchers could fill the void. They have not because the research culture, even among laboratory statisticians and worldly economists, continues to value publication in academic journals over solving an agency's data problems.

4. THE ORNL DATA QUALITY CONTROL CYCLE

My basic problem with the ORNL data quality control cycle (Fig. 2) is that it assumes the old system still works. It does not. Most of the highly trained government analysts are gone, and technically accomplished managers are viewed with suspicion. If a quality-control approach is to have any chance of success, it must tip the balance in favor of data quality research and application. None of us can do a thing about the economics of information, federal budget deficits, or the civil service. We can do a great deal about research tactics.

The ORNL quality cycle is totally sealed off from statistical agencies, much like a physician who refuses to deal with the sick. Where are the agency managers and technical staff? How do researchers know what the agency needs, what the agency will comprehend? How do researchers learn what is possible for the agency? Where does education, trial and error, and regrouping take place?

To progress, researchers must become much more aggressive and far less prudish in dealing with people working in agencies. At a minimum, the

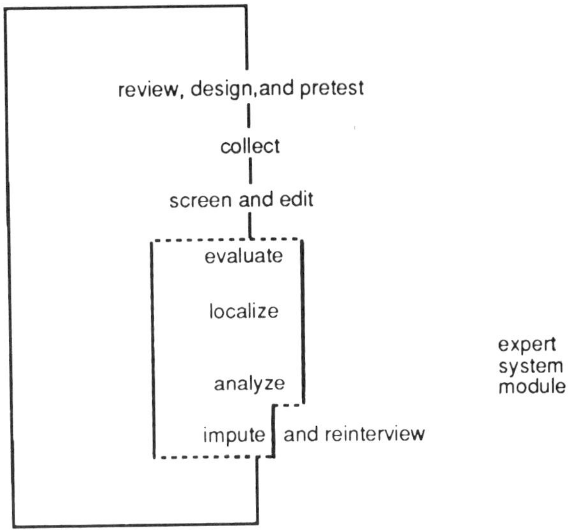

Figure 2. ORNL data quality control cycle.

data quality cycle should have a large block called marketing, education, and hand-holding. Activities within this block should include the following:

1. *Gather intelligence* on the needs and pressures facing potential clients.
2. *Educate researchers* about the data problems as the agency understands them.
3. *Educate agency technical staff and managers* about what is known, what is knowable at low risk, and what is knowable at high risk.
4. Provide *rewards, risk sharing, and commiseration.* Commiseration should be capitalized because research projects often fail. That is why it is called research. Do not simply abandon the disappointed sponsor who has to explain to his or her bosses why there is nothing to show for the quarter million dollars he or she talked them out of earlier.

Such a simple, straightforward accommodation to the newer realities of life in federal statistical agencies would do much to improve the prospects for data quality. It would, however, require a major change in the attitudes of many laboratory and university researchers. Scientists have long derided the marketers—just think about Carl Sagan's treatment or Andy Loebl's. Those days would have to end.

5. RESEARCH TOPICS

I promised the chairman a few researchable topics. I will keep that promise.

From my point of view, the Panel on Incomplete Data's suggestion to develop standard software is an excellent one. Our data-processing, reporting, auditing, and training burdens would be substantially reduced by a library of reliable subroutines for survey operations.

We are also interested in "electronic supervisors" to monitor the work of contractor survey staff. We have discovered excessive variation in how edit flags are resolved and how callbacks and imputations are actually performed. Costs also vary widely. We are in the preliminary stages of trying to use personal computers and artificial-intelligence shells to discipline these decisions.

Over the years, our audits have uncovered instances where the computations made are not the calculations that the operators and the documentation describe. The programs run and there are no error messages to indicate the condition. I would be interested in any statistical, mathematical, or software-engineering procedures that would alert us to these subtle discrepancies. Since our largest programs are small compared to military applications, I still hope for something more reliable than the human sleuths we use now.

EIA's source and reliability statements about nonsampling error leave me cold. We have extensive, though unassembled, respondent-level data on edits, imputations, revisions, final resolutions, and comparison with related series. We are not using the data to tell the reader how accurate this month's estimate is likely to be. I am sure that there are good summary measures. We just need to define them and try them.

I also hope that the research and controversy surrounding the census under count might have yielded ways to test the completeness of EIA's frames. I have not had time to look.

6. CONCLUSIONS

The biggest obstacle to progress in data quality is not the lack of worthwhile research topics. The biggest problem is that the environment for developing ideas has changed and some of our ways of thinking about research have not. I hope my remarks will assist researchers, especially those at Oak Ridge National Laboratory, as they develop the ideas that let us take the next steps toward better data.

APPENDIX

Crosswalk from Table 2 to the Panel's Recommendations

Chart Heading	Recommendation Number
Get the data	1
Good collection, form and data...	2, 3, 10, 11, 12, 13
Measure and manage...	4, 5, 6, 7
Design for nonresponse	8, 9
Analyze	
Effect...	14, 15, 21, 22
Alternative...	16, 17, 18, 19, 20
Documentation	23, 24

Crosswalk from Table 3 to the Panel's Recommendations

Chart Heading	Research Recommendation Number
Standard software	1
Investigate nonresponse mechanism	3, 4, 5, 10
Evaluate imputation schemes	2, 6, 7
Develop consistent terminology	9
Determine C/E...	11
Investigate effect...	8

3

Reflections on Validation and Quality Assessment of FPC Form 4 Data

GUNAR E. LIEPINS Energy Division, Oak Ridge National Laboratory, Oak Ridge, Tennessee

1. INTRODUCTION

The FPC Form 4 is used to collect data annually about some 6000 power plants in the United States (virtually a census of all the larger power plants). Required submissions include location, prime mover, capacity, generation, consumption, fuel stocks, and primary and alternate fuels. In 1979, the Energy Information Agency (EIA) of DOE charged Oak Ridge National Laboratory (ORNL) with validating the Form 4 data for the years 1972 to 1978; that is, determining its quality, accuracy, usefulness, respondent burden, and overlap with other data-collection efforts. This chapter briefly describes what was done technically, places those activities in the framework of the proposed data quality-control cycle, and identifies how the study could have been directed more purposefully.

2. QUALITY ISSUES

As has been articulated at the workshop repeatedly and effectively, data-quality issues encompass more than just technical analysis and statistical manipulations. Quality improvement requires communication, commitment, and educational efforts. Moreover, quality can be measured only relative to use. For example, data that are adequate to determine whether power usage has increased over a historical period could well be inadequate to support a more detailed analysis of power usage patterns or effects of fuel disruptions and cost variations.

Unfortunately, the federal mandate for the collection of the Form 4 data failed to make explicit and precise the uses that those data should support; presumably, the intended purpose was to support power requirement and contingency planning. Thus, data adequacy and quality could be evaluated only in terms of the "most important and likely" hypothesized uses. The most likely of these seems to be the assessment of the nation's ability to provide different levels of power (and of the cost of that power) under various fuel-disruption scenarios. This would include determination of capacity, efficiency, and possible alternative fuel usage. A second use of the data might be to determine environmental impacts of power generation.

Despite the lack of a clear standard and objective against which to measure the quality of the FPC Form 4 data, the ORNL evaluation began with a careful review of the data-collection instrument and screening of the data (Liepins et al., 1979). (Without such an analysis, even with an explicit purpose for the data, no quality assessment is possible.) The first step of that analysis corresponds to the first step of the quality cycle: review. The review elicited the following observations:

1. No clear definition of "plant" was given.
2. The data system was not designed to generate a unique identifier for each plant. Duplicate records could be present in the data.
3. The data format made it impossible to track transfers of plants from one company to another. The time evolution of power plants would be difficult to follow.
4. Blanks were used in the status fields for operating plants. A plant with a missing status field might be interpreted as operating.
5. A plant code of "000" was entered for both aggregated plants and those for which information was missing. The two cases could not be distinguished.

The screening step confirmed some of the problems raised by the review and identified other problems:

1. Sixteen duplicate records were found in the database. (More exactly, seven duplicates, three triplicates, and one quadruplicate were found.)

2. Two plants were reportedly added and retired in the same year. By itself, this does not represent an inconsistency. However, it did seem to cause confusion on the part of the reporting company about how to report the required information.
3. Plants retired during the reporting year were stated to have zero capacity. In truth, they probably continued to have generating capacity, but the individual companies had chosen not to use them.
4. Changes in capacity were reported predominantly in January and December and represented approximately 300 percent and 150 percent, respectively, of median reported capacity changes over the calendar year. Presumably, this reflected the bookkeeping policies at the individual companies and not the actual timing of capacity changes.
5. Multivariate outlier detection identified several records as being unusual. In one case, an apparent keypunching error resulted in reported consumption two orders of magnitude greater than actual consumption. Another plant was determined to have reported fuel oil usage in gallons instead of in barrels. Several of the outlier plants proved to be cogeneration plants, a feature not addressed by Form 4.
6. Over the years, capacity was reported to be zero for between 1 percent and 3 percent of the records; generation was reported to be zero for between 3 percent and 5 percent; stocks were reported to be zero for between 4 percent and 8 percent; and consumption was reported to be zero for 8 percent. Part of the difficulty can be traced to the inclusion of FPC Form 12 data in the FPC Form 4 database. For the former, it is expected that company code would not be reported; zeros were used both for missing values and to report a quantity of zero. The majority of the records were complete in the important fields; 85 to 88 percent of the records had no missing capacity, generation, stocks, or consumption fields. The application of edits (or consistency constraints) generally yielded failure rates on the order of tenths or hundredths of a percent for each edit. Exceptions included the constraint that generation not exceed capacity, which was violated by approximately 3 to 10 percent of the records (depending on the year). In truth, capacity was reported as nameplate capacity and not as actual capacity. Other frequently violated edits were of the same type; generation exceeded stated limits for the particular fuel type. Error localization and imputation were not performed on the data.

What is the quantitative significance of these statistics and findings? What, if anything, can be said regarding quality independent of data use?

Unfortunately, without additional analysis into the independence or dependence of the various errors and omissions, their additivity or redundancy, and follow-up on a sample of records that passed the consistency constraints, only rough qualitative bounds can be given regarding data accuracy. Even for rough bounds, not all of the required statistics were compiled and published in the validation study. For example, it would have been useful to know the following: For the plants whose generation exceeded capacity, what is an estimate of the understated capacity? What percentage of records were complete, failed no edits, and were not flagged as suspicious by any of the analyses? What percentage of total generation was associated with those complete, consistent, nonsuspicious plants?

The analyses performed and omissions made by the validation study provide lessons for future quality assessment. First, a combination of careful review of the data-collection instrument, exploratory data analysis, outlier detection, consistency checking, and selected follow-up provides a paradigm for determining data irregularities and hidden phenomena. Second, even though data validation has its exploratory components, it is useful to identify the critical data variables at the outset of the validation, plan the validation to yield complementary evidence on the accuracy of those variables, and produce rough bounds on their estimated accuracy.

Given that the FPC Form 4 study focused more on what individual technical analyses might uncover than on pulling together the results of those analyses, can anything be said about the quality of the FPC Form 4 data in terms of their hypothesized uses? Unfortunately, the answer is yes only for structural questions such as "Can emissions be calculated indirectly from the data?" and a speculative "maybe" for quantitative questions related to accuracy of derivable estimates, such as "What is the 90 percent symmetric confidence interval for power generation in the face of a 20 percent fuel oil shortfall?"

Perhaps it is not self-evident that the study's failure to address quantitative accuracy-related questions is only partially attributable to the failure to integrate the various analyses and follow them to their natural conclusions. A validation study or quality analysis can address only a small fraction of the possible questions about data quality and adequacy; such studies and analyses need to be driven by the explicit data requirements. Open-ended speculation such as "What might these data be useful for?" is, at best, a rationalization after the fact. Data collection should be driven by a need for these data. Needs should not be invented to justify the data. Data for which there is no need should not be collected. The lack of a need for the data represents a legitimate validation finding and should result in the recommendation that the data collection be discontinued.

3. CONCLUSION

The review stage of the FPC Form 4 study identified a variety of structural shortcomings in the format of the collected data (lack of unique record identifiers and ambiguity of coding). Resolution of these structural weaknesses would improve the quality of data collected in future years. Outlier detection methods and exploratory data analysis revealed previously unknown features of the data. An important lesson for future validation studies is that such studies are most meaningful when they are directed at determining whether the data in question adequately meet the real needs of the data users. The gap between the informational needs and the data coverage (including accuracy) reflects the shortcomings in data quality and suggests areas for data improvement.

REFERENCES

Liepins, G. E., C. L. Begovich, M. R. Chernick, P. M. Dizillo-Benoit, V. E. Kane, and D. J. Pack (1979). *Data Analytic Tools in Support of Data Validation*, Report to the Energy Information Administration, Contract No. W-7045-eng-26, Oak Ridge National Laboratory.

4

Data Quality Control at United States Fidelity and Guaranty Company

NANCY J. LePAGE United States Fidelity and Guaranty Company, Baltimore, Maryland

1. INTRODUCTION

The United States Fidelity and Guaranty Company, herein called USF&G, is a multiline insurance company principally engaged in the business of writing property/casualty insurance, including personal, commercial, and fidelity-surety.

The home office is located in Baltimore, Maryland. Fifty-four branch offices and numerous suboffices are located throughout the United States. The principal distribution of our products is the independent agency system. Approximately 7000 independent insurance agencies represent USF&G.

USF&G has a long history of dedication to quality and has developed a corporate culture based on hard work, teamwork, and high quality standards. In mid-1977, there was a call by the insurance industry for increased quality control/quality assurance. In response, USF&G reinforced its commitment to quality by establishing a Quality Assurance Department in September 1977. While the operating philosophy of our department encompasses all aspects of service quality, ensuring the accuracy of our statistical data is our primary focus.

The need for accurate data in the insurance industry is crucial. Inaccurate premium or loss data can adversely affect company pricing, marketing, underwriting, loss reserves, and bottom-line operating results. Underwriting and marketing decisions based on such data can lead to continued commitment to unprofitable coverages or risk concentrations.

Inaccurate data also affect a company's relationships with the public, insurance regulators, and insurance agents. Loss of goodwill and confidence in the company can result.

In response to the need for accurate data in rate-making, as well as to regulatory pressure to improve data submissions, the Quality Assurance Department developed a company-wide quality-control program in 1979. The institution of a quality-control function allows management to establish accountability for the quality of data and to improve the acceptability and credibility of those data.

The primary objective of USF&G's quality-control program is to ensure that our company's statistical information accurately reflects the true composition of risks we insure and, in a timely manner, the premiums and losses associated with those risks. Our quality-control program includes four elements: prevention, control, assurance, and audit.

Prevention involves detecting and correcting problems before our products and/or services enter the customer service cycle. By continuously monitoring the accuracy of data at all stages of processing, errors can be prevented.

Control is established by safeguarding against data manipulation. Each time data are transferred, changed, translated, or reproduced, the opportunity for data manipulation is present. Examples of controls include computer reasonability edits and independent reviews of work.

Assurance involves guaranteeing that our products, processes, programs, services, and people are performing in accordance with established company and industry standards. Compiling and analyzing data-quality problems and offering recommendations to improve processes are part of the assurance function. Audits indicate whether or not the objectives of our prevention, control, and assurance efforts are being achieved. The audit function is the main focus of this chapter.

The audit is an independent assessment of adherence to predetermined standards. The purpose of the audit is to objectively measure and report compliance with requirements. Due to the inherent nature of the auditing process, this is sometimes perceived as an adversarial confrontation between those being audited and the auditors. This perception is not accurate, since both sides have the same objective: ensuring proper control over data processing. If problems are not identified and corrected, they can lead to poor management decisions and a loss of confidence in the data-handling process, both of which create adverse effects on the industry and the consumer. A properly conducted audit provides management with useful information necessary to make sound business decisions and is, therefore, an absolute requisite for quality service, however unsettling the audit process may be.

Our Quality Assurance Department routinely conducts three types of audits: the system audit, the statistical audit, and the procedural audit. I will begin by discussing the system audit.

2. THE SYSTEM AUDIT

Important changes in recent years have affected management's need for information to plan, evaluate, and control the operations of our business. Increasingly, management requires more comprehensive information to make timely decisions. This growth in the need for management information has been paralleled by the growth of data processing. Management at all levels has become dependent on data processing and, consequently, more concerned about the continuing accuracy and completeness of data-processing results. With the introduction of new technology and with this greater dependence on data processing, the need for the auditing function is more apparent than ever.

Insufficient control is a frequently reported concern of management. This concern is based on the increasing dependence on computer systems and the reality that the opportunity for errors and omissions increases as computer systems become more autonomous, comprehensive, and complex. Auditors are needed to determine the reliability of management data developed within the organization.

Our Quality Assurance Department conducts system audits to ensure that our company's computer systems are operating as planned and as required. System audits enable us to identify and act on problems, or potential problems, related to the collection and reporting of company data. Our objective is to ensure that our company's data are in compliance with established standards for accuracy.

Standards for data accuracy have been developed for the insurance industry by such organizations as the Insurance Services Office (ISO). The ISO is a national advisory rating organization that gathers statistical information from insurers, compiles the data, and then disseminates the data to regulators and insurers. These data are used as the basis for ratemaking decisions; therefore, the level of data accuracy is critical. The ISO established error-tolerance levels for each line of business by conducting studies to determine what degree of accuracy must be achieved to maintain the validity of the data they receive. These tolerance levels are the industry standards by which statistical data accuracy is measured.

We have developed standardized procedures for planning and conducting system audits of our company's computer systems. In developing our procedures, we relied heavily upon the ISO's *Quality of Data Audit Guide* (1978) as a reference. This guide provides step-by-step procedures for conducting a comprehensive audit of data quality. It was developed by the ISO to assist companies in ensuring the accuracy, completeness, and timeliness of their data. Guidelines for the following are provided: audit preparation; reconciliation of statistical and financial data; sampling criteria; sampling and testing; error detection and correction; analysis, recommendations, and reporting; and follow-up.

At USF&G, we approach the system audit in five phases: preliminary research, design, data gathering and analysis, final report, and follow-up. I will discuss each of these phases separately.

2.1 Phase I: Preliminary Research

During the preliminary research phase, we identify which system is to be audited and then document the audit's feasibility. For example, for the audit to be successfully conducted, the necessary resources must be available. Qualified personnel are needed to conduct the audit, and the owners and/or users of the system are needed to provide support. Additionally, all materials and supplies necessary to conduct the audit must be accessible.

A system audit may result from a request by the owners or users of a system, or the Quality Assurance Department may initiate an audit. For example, in 1986, the Quality Assurance Department initiated an audit of our new Personal Lines Dwelling System. The purpose of this audit was to verify the accuracy of computer-generated coding and ensure that the system was processing dwelling policy data correctly. The system was audited during its initial phase of implementation in order to detect problems and initiate corrective action before the system went countrywide. At the time of the audit, 27 of our branches were using the system. The Dwelling System is a policy-writing system and interfaces with three other company systems; therefore, it was imperative that the data being produced by the system were reliable.

If, as in this example, we initiate the audit, our first step is to set up a meeting with the owners or users of the system to discuss our reasons for selecting their system and to talk about our audit plans. It is critical that the owners' support for the audit be obtained at this early stage. They are the most familiar with the system and will be best able to provide guidance in determining the critical areas to be audited. Additionally, they may have concerns that can help determine the focus of the audit. Also, it is the system's owners who will be responsible for implementing corrective action to resolve any problems detected during the audit; therefore, the effectiveness of the audit depends on the cooperation of all parties involved.

In most cases, the focus of our audit is on compliance, as opposed to noncompliance. Traditionally, audits have been focused on finding nonconformities after they have been produced. This involves rework, rehandling, and reinspection of nonconforming products.

By focusing on compliance, systems are monitored to determine when adjustments are needed to maintain stability or to respond to the inherent variability of business needs. By focusing on defect prevention rather than defect detection, when problems are detected, corrective action can be taken before a nonconforming item is produced.

By assuming that requirements are being conscientiously followed and by communicating this assumption to those being audited, fear of the auditors is allayed. This in turn helps build a relationship of trust, which will contribute to the audit's success.

Once a system is selected, a data-flow diagram is prepared and a system description narrative is developed. The data-flow diagram provides a graphic description of the system's overall structure and its relationship to other systems. Inputs, outputs, and major functions of the system are illustrated. For example, Figure 1 indicates the data flow of the Personal Lines Dwelling System.

The system description provides nontechnical information on the system and helps us to understand the system we are planning to audit. The description includes such information as number and location of users; process frequency; interfaced systems; system inputs and outputs; and existing input, output, and on-line processing controls.

2.2 Phase II: Design

Phase II, design, is intended to ensure that the audit will be orderly and comprehensive. Human resource needs are established. Normally, two or three quality-assurance analysts will be involved in the audit on a full-time basis, and limited participation of the system's owners will be required. Depending on the nature of the audit, additional resources may be

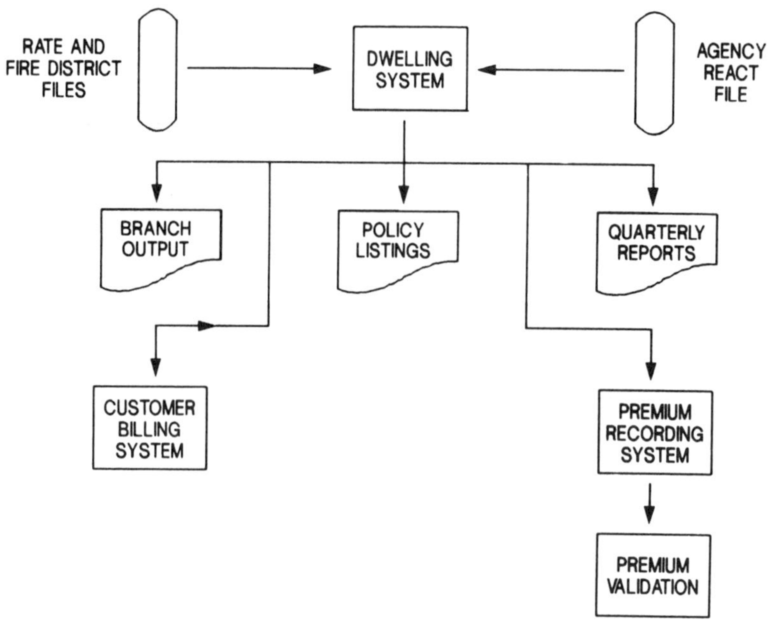

Figure 1. System flow.

required. For example, during the Dwelling System audit, we called upon the services of the Quality Review Unit to assist us in reviewing statistical coding for accuracy. The Quality Review Unit is a part of the Quality Assurance Department with responsibility to conduct statistical coding reviews. I will discuss this unit in more detail later on.

Next, a Gantt chart is prepared. A Gantt chart is a project-management tool that is used for scheduling. This chart lists all activities that will occur during the course of the audit and sets time frames for each activity. For example, Figure 2 illustrates a portion of the Gantt chart used for our Dwelling System audit. Only the major activities and their time frames are illustrated in this figure. On the actual Gantt chart prepared for this audit, all the activities that would occur under each major task were also listed. Planned duration for a system audit is usually six months, and by maintaining and updating a Gantt chart, we are able to monitor our progress.

It is during this phase that we determine which segments of the system are to be audited and what types of transactions are to be

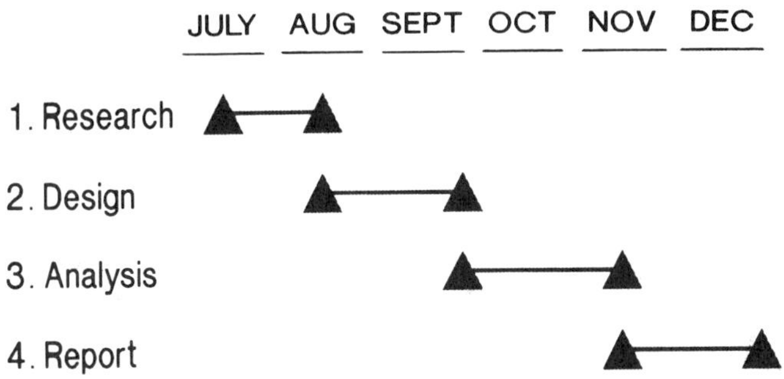

Figure 2. Gantt chart.

examined. The statistical data elements to be reviewed for accuracy are selected. For example, Table 1 illustrates a partial list of the codes we selected to verify for the Dwelling System audit. Next, the sources of information that will be used to measure accuracy are secured. At this stage, input from the system's owners and users is very valuable.

Next, we determine which point of the processing will be our sampling point and what types of computer listings and printouts will be needed to thoroughly audit the system. A determination is then made regarding what specific source documents will be necessary for the audit.

The next step is to determine our sample size. While it would be ideal to be able to detect all system errors that have occurred, it is not cost effective to inspect every item processed by the system. To determine a valid and reliable sample size, we assess four factors: the population size, the estimated error ratio, the reliability range, and the confidence level.

The population size is the number of items from which the sample is to be drawn. The estimated error ratio is the error percent that can be

Table 1 Data Elements

- Line
- Subline
- State
- Territory
- Premium
- Protection class

tolerated or expected in the population. In general, we use an estimated error ratio of 5 percent. The reliability range is the degree of precision desired in the sample results, and we normally use a factor of plus or minus 2 percent. This means that the error ratio for the entire population is expected to be between 3 and 7 percent. The confidence level is the degree of risk that the error ratio found in your sample accurately represents the entire population. We use a confidence level of 95 percent. Given the values we assign to these four factors, we are then able to determine a valid sample size using statistical sampling tables, such as those found in the *Handbook of Sampling for Auditing and Reference Accounting* (Arkin, 1974). The next step is to develop the specific procedures that we will use to review sample items. We then design control logs to record the inspection results.

At this point, the quality-assurance analysts may need formal training on how to properly evaluate the data accuracy of the system and interpret the data-entry requirements. The system's owners are the best qualified to provide this training; therefore, we request their assistance.

2.3 Phase III: Data Gathering and Analysis

The third phase of the audit involves the actual data gathering, review, and analysis. This phase encompasses the tasks necessary to evaluate statistical data accuracy in accordance with documented processing procedures and to draw verifiable conclusions.

The items to be included in our sample are selected. To select our sample for the Dwelling System audit, our Personal Lines User Support Department provided a list of dwelling policies processed by the system for the time period in which we were interested. The sampled policies were randomly selected from this listing. The source documents for our sample were requested and obtained from the branch offices, and the applicable computer listings and printouts were obtained from the User Support Department.

The source documents and the computer-generated coding are then compared and reviewed for accuracy. The source documents are compared to the initial system record, and all input documents are compared to all output records. Errors are identified and documented.

Next, we determine the source of each error. Discrepancies are traced back through the intermediate system printouts to determine where the discrepancy originated. The source of the error may turn out to be the branch office. Coders in our branch offices may not be coding the item correctly. Or, we may find the system itself is the source of the error.

Once the source of each error is established, further research is conducted to determine the actual cause of each error. For example, incomplete coding instructions may be causing branch-office errors. Unreliable edits may be causing system errors.

If the system is causing errors, further research is needed to determine the specific origin of the errors. For example, the system may be producing inaccurate data due to a flaw in the original design or development of the system. Or, errors may be a result of the system maintenance procedures being applied.

Once the entire sample has been reviewed, discrepancies are summarized by source, type of error, and frequency of occurrence. For example, Table 2 illustrates the errors detected during the Dwelling System audit.

The next step is to meet with the system's owners and/or users to verify our findings. Where problems are found to exist, alternatives for corrective action are explored. Ideally, the auditors and the system's owners are able to come to an agreement on the corrective action to be taken.

Next, we determine the impact of the errors found. Who do they affect and how severely? Errors undeniably create increased costs for the company, both the costs of actually having to reprocess the work and the costs associated with having to delay further production until the reprocessing has taken place.

In most cases, we are able to quantify, in dollar amounts, the impact of inaccurate data on the company. This quality-cost report serves several purposes. It illustrates an actual dollar cost to the company; therefore, it is

Table 2 Errors Detected by Dwelling System Audit

Cause	Error Type	Number of Errors
System	Protection	11
System	Type phrase	5
System	Construction	1
Branch	Zip code	1
Branch	Number of families	1
		19

likely to gain management's attention. It clearly illustrates where cost-saving opportunities exist within the company. It relays our findings in concrete, measurable terms. It provides a justifiable way to determine whether or not immediate corrective action is necessary. Additionally, cost figures provide support for the project team's recommendations, which appear in the final report.

2.4 Phase IV: Final Report

The fourth phase of our system audit is the final report. The report consists of seven sections: Executive Summary, Background, Methodology, Findings, Cost Analysis, Conclusion, and Recommendations. The final report is sent to the manager of the affected department.

Before release of the final report, an initial draft is prepared for review and comment by all participating departments. The purpose of the draft is to enable each department to add its clarifications or corrections to our findings and make a commitment to addressing the report's recommendations. This is an important step and has several advantages. By reviewing the initial report with the affected departments, the report becomes a group report instead of an evaluation by the "corporate policemen." We then have the opportunity in the final report to advise that some of the problems found have already been addressed and to indicate the status of those recommendations that will be addressed.

As stated previously, the focus of the audit is on compliance, as opposed to noncompliance, and this is reflected in the final report as well. The report provides appropriate recognition when processes and systems are functioning properly. The report not only highlights the deficiencies and weaknesses found, but also discusses the positive points.

2.5 Phase V: Follow-up

The final phase is follow-up. We begin follow-up within six months of the completion of an audit. The follow-up is conducted to confirm and evaluate the corrective action taken as a result of the audit findings. A minimum follow-up is appropriate if the audit findings are such that an independent verification of the problems found is not required or would not be cost justified. However, if a large number of errors or error types were identified and an independent verification of correction is cost justified, a mini-system review will be conducted. In the event that the original audit identified major problems traceable to the system's initial design, development, or testing processes, then a System Assurance Review will be conducted.

A system-assurance review is a management-oriented evaluation of a system's adherence to its original business requirements, design specifications, development and documentation standards, testing methodology, and project monitors (changes to the system after its initial design and development).

In the case of our Dwelling System audit, we found that errors were occurring as a result of the original data-processing programming. The recommended programming changes we made in our final report were implemented; in this instance, only minimal follow-up was required.

This concludes my overview of our process to verify the accuracy of our computer data. But not all data-accuracy problems are the result of computer data processing. In monitoring the accuracy of USF&G's statistical data, we have found that a high percentage of our errors originate within the branch office key-entry process. For this reason, we also conduct Statistical Audits to verify the accuracy of manual coding.

3. THE STATISTICAL AUDIT

USF&G has long been concerned with the quality of manual coding. I mentioned the Quality Review Unit earlier, and now I would like to discuss the purpose of this unit.

In 1961, our Data Processing Department started a unit known as Quality Control. This unit is responsible for conducting statistical reviews of branch-office coding. The name of the group was changed from Quality Control to the Quality Review Unit in 1981. This unit is headed by the supervisor of quality assurance and is staffed by four full-time expert coders and one clerical assistant.

The Quality Review Unit (QRU) contributes to our commitment to improving the quality of coding produced by our branch offices. This unit is responsible for performing independent statistical coding reviews of our branch-office premium and loss coding. This task requires a detailed review of a sample of branch-office policies and corresponding coding.

Since this unit's inception, it has been responsible for performing statistical audits of our Casualty and Property lines of business. As a result of a system audit we conducted on our Private Passenger Automobile Claim System in 1981, loss coding is now included in the statistical audit process. Our system audit had shown that the error level of our branch-office claim coding was unacceptable. Therefore, we currently are conducting statistical audits on branch-office coding of manually issued Commercial Property and Casualty business, manually issued Personal Property and Casualty business, and Claim Transmittals.

Two times a year, manual coding from each of our branch offices is reviewed for accuracy. For a selected period of time, a branch will forward copies of all items it processes to the Quality Review Unit. Using statistical sampling tables, the QRU will select a sample of files from each line of business to recode. Items subject to review include new business, renewals, endorsements, cancellations, and transmittals.

Control sheets are used to record the results of the audit. The specific coding errors detected are listed, and the correct codes are indicated. Separate control sheets are completed for each line of business. All discrepancies are noted in a report returned to the branch office. This report identifies the number and types of coding errors that were made. It also indicates the overall error percentage detected by the QRU in each line of business.

Statistical audits enable us to determine and monitor the data-accuracy levels of each of our branch-office departments and each line of business and to initiate corrective action when necessary. We use the results of the audits as tools for improving the data accuracy of our company. Each of the five analysts in the Quality Assurance Department is assigned a certain number of branches to work with and assist in improving data quality. Once the statistical audit of a branch office is completed by the QRU, a copy of the report is given to the assigned analyst. The analysis performed by the analyst serves to highlight, to the particular department audited, the areas of coding in which immediate corrective action needs to be taken to prevent recurrence of problems.

One very common method of analysis is Pareto Analysis. Pareto Analysis is sometimes called the 80/20 *principle* because it is based on the assumption that a large portion of all errors, about 80 percent, is caused by only 20 percent of the possible contributors. The major error types are easily identified through Pareto Analysis. Errors are listed by frequency of occurrence. For example, Table 3 is a Pareto Analysis of the Commercial Auto line of business. Thorough analysis of the major, or most frequent, errors will help to uncover the underlying causes of these errors. Pareto Analysis enables one to establish priorities for eliminating errors. We are able to focus our efforts on the areas where the greatest improvement in quality will be realized.

Another benefit of the statistical audits is that they allow us to monitor the accuracy of coding over time. For example, Figure 3 shows a

Table 3 Pareto Analysis of Commercial Auto Line

Error Types	Total Errors	Percentage of Total
Exposure	20	32
Deductible	18	29
Line	10	16
Subline	9	15
Class	5	8
TOTAL	62	100

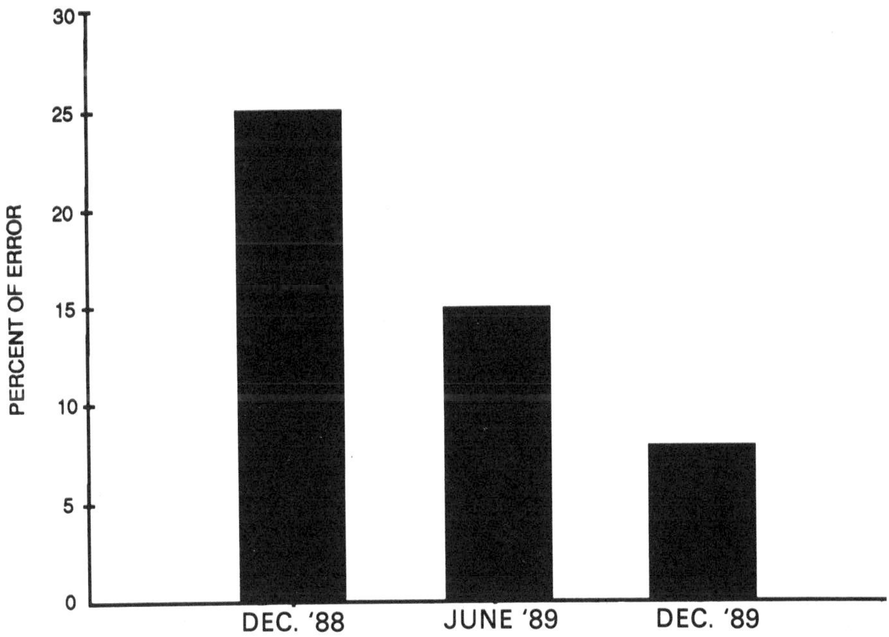

Figure 3. Personal lines.

trend analysis of a Personal Lines Department. This graph illustrates a continued improvement in the quality of statistical data over the past year and a half.

Analysis methods such as these enable us to help our branch offices improve their coding accuracy and thereby improve the quality and reliability of our company's statistical data. Our statistical audits also point out areas where errors have been occurring, but have not been detected within the branch. Therefore, the audits serve to clarify misunderstandings of coding instructions.

Besides providing us with insight into the coding problems of each particular branch office, the audits also enable us to determine leading country-wide problems. We have set up a database to record audit results. Each month, the number of times each code is found to be in error and the number of branches that contributed to the errors that particular month are entered into the database. This information is then summarized and analyzed by the Quality Assurance Department at the end of each year.

Using Pareto Analysis, we are able to determine the leading country-wide problems, by line of business. For example, Table 4 illustrates the most frequent error types detected on the Commercial General Liability line of business during 1986.

After the Pareto Analysis is prepared for each line of business, the next step is to research causes of the predominant error types. We go back to the source documents and review the miscoded items to gain insight into the actual causes of the errors.

The results of the analysis of our database information can be used to identify areas where the need for clarified coding instructions is apparent.

Table 4 Error Types Detected on Commercial General Liability Line

Code	Number of Errors	Percentage of Total
Exposure	516	20.2
1. Bodily injury/ property damage	339	13.2
Deductible	219	8.5
Class	174	6.8
2. Rate identification	171	6.6
3. Rate modification	165	6.4

Results can also be used to advise our branches of the leading country-wide error types and provide them with some insight as to the causes of these errors. Awareness of problems and their likely causes is the first step toward a reduction of errors and an increase in data quality.

The statistical audits conducted by the Quality Review Unit also enable us to compare country-wide accuracy levels over time. We have experienced a gradual, but substantial, decrease in our country-wide error percentages since the statistical audits were first implemented.

As our error percentages have decreased, so has our cost associated with finding and correcting statistical coding errors. In a Quality Cost study we conducted in 1985, we determined that it costs an average of $2.31 to correct each error made by our branch office Personal Lines, Commercial Lines, and Claim Departments. Therefore, the reduction in error percentages has resulted in very substantial cost savings to our company.

Based on the improvement we have seen in our data quality over the past several years, we feel that our statistical audits have had a very positive effect on the accuracy and reliability of our company's statistical data. We plan to continue this audit process as long as USF&G continues to generate manual coding.

4. THE PROCEDURAL AUDIT

I would like to briefly touch upon the third type of audit we conduct—the procedural audit. The Quality Assurance Department has implemented a quality-control program in each of our branch-office Claim, Commercial Lines, Insurance Engineering and Audit (IEA), Fidelity-Surety, Personal Lines, and Service Departments. The purpose of our procedural audit is to verify that each branch-office department is operating its quality-control program in accordance with the requirements of the program, as stated in the Quality Control Manual.

For example, the Quality Control Manual states that 20 percent of the coding produced in our branch offices must be inspected for quality control. A sample of each individual's work must be included in the inspection, and the inspection results must be documented on quality-control forms. Through our procedural audit, we are able to verify compliance or identify noncompliance with these required procedures.

Procedural audits are conducted by visits to the branch offices by our quality-assurance analysts. In addition to verifying compliance with procedures, we also focus on the areas of analysis and corrective action.

Only if our branch-office personnel are making use of the data generated by the quality-control program to improve the quality of their coding will the objectives of the program be met and the benefits realized.

These visits allow us an opportunity to reinforce the program philosophies, as well as to fine-tune the controls that are in place. Additionally, we feel that these audits demonstrate to branch-office personnel the interest and concern of USF&G management in ensuring and maintaining the quality of our statistical data. The quality audit also helps to build and maintain employee interest in quality.

5. CONCLUSION

We are all aware of the detrimental consequences of inaccurate data. Data reported to statistical agents have a direct impact on our industry's final consumer product cost. Additionally, the accuracy of data submittals impacts on the reputation of the insurer.

USF&G's goal is to produce accurate and timely data to ensure that our statistical information properly reflects the exposure we write. It is our goal to provide quality data and quality service at the lowest possible cost. We are striving to achieve this goal through conscientious adherence to a coordinated and documented quality-control program.

Audits are an essential part of our quality-control program. Audits enable us to compare performance against established standards and initiate corrective action where it is needed to improve the accuracy of our statistical data. System audits, in particular, enable us to ensure that our company's computer systems are operating as planned and as required. By conducting our five-phase system audit, we are able to assess the adequacy of our controls, identify and correct problem areas, and provide assurance that our systems are producing accurate data.

As a result of our quality-control program, we have had a significant reduction in our statistical data-error rates, and as a result of our quality audits, we have gained a greater confidence in our overall data integrity. But the challenge does not stop there. We will continue to compare actual performance to desired performance to ensure the continuous production of quality products. And, as a final control of our data integrity, we will periodically review our internal standards for performance to ensure that they are adequate to meet the changing needs of today and tomorrow.

REFERENCES

Arkin, Herbert (1974). *Handbook of Sampling for Auditing and Accounting*, McGraw-Hill, New York.

Insurance Services Office (1978). *Quality of Data Audit Guide*, Insurance Services Office, New York.

5

Quality Control for a Shared Multidisciplinary Database*

NANCY FLOURNOY The American University, Washington, D.C.

LEONARD B. HEARNE George Mason University, Fairfax, Virginia

1. INTRODUCTION

A common information model for business, industry, and medicine includes a database system and system users who need timely access to accurate information. One of the outstanding problems associated with implementing this model is maintaining high-quality data in the database without resorting to policies and procedures that are external to the database system. The authors have developed a methodology that promotes quality data within a database system. To illustrate key points in this methodology, examples are drawn from our experience in developing and maintaining a database system at the Fred Hutchinson Cancer Research Center (FHCRC).

Central to the research effort is a shared database of patient demographic, clinical, and laboratory data. An interdisciplinary team of clinical health-care providers and laboratory research scientist, distributed over a wide geographic region, contribute to and retrieve data from this shared database system (Figure 1).

*This work was supported in part by National Cancer Institute grant CA18029 while the authors were at the Fred Hutchinson Cancer Research Center, Seattle, Washington, between 1975 and 1986.

Laboratories

Virology Biochemistry Microbiology Immunology

Hematology Pharmacology ···

Facilities Management

CORE DATA

Administrative Services

Clinicians Nutritional Scientists

Nurses Pharmacists Social Workers

Expert Clinical Opinion and Observations

Figure 1. Multiple disciplines and sources of data.

The authors have developed an architecture and a methodology for incremental database system development at the FHCRC that promotes data quality. Central to the system design methodology is the creation and promotion of positive incentives for system users. Incentives are tailored to classes of system users, depending on the perspective from which the user views the database system. The users' views are defined and classified from the interrelationships between their needs for accessing, manipulating, and making decisions based on the data in the system.

We begin by introducing the clinical and research environment at the FHCRC as a multidisciplinary model. Then, we introduce the classes of system users, followed by the perspectives with which they view data, and the database system. We conclude with guidelines for designing and implementing database systems that promote quality data.

2. THE MULTIDISCIPLINARY RESEARCH ENVIRONMENT

Research is increasingly performed by multidisciplinary teams of scientists, bringing their varied training and expertise to bear on common scientific objectives. For example, the common objective of the research at the FHCRC is to find cures for cancer. Only 10 years ago, clinical cancer research was primarily the purview of two disciplines: chemotherapy and radiotherapy. Thus, a rather limited number of specialties within the medical sciences was involved. The single measure of scientific outcome was simply how long patients remain free of disease following treatment.

The FHCRC in Seattle began admitting patients with hematologic malignancies (e.g., leukemia) in 1975. Since that time, cancer research has evolved to include investigations into the operation of the human immune system: on the cellular level with cell biologists, microbiologists, and virologists and on the molecular level with biochemists, biophysicists, and pharmacologists. The objective of these studies is to establish how the human organism recognizes inappropriate conditions within itself; how it mounts a defense against these conditions; and what clinical, cellular, and biochemical measures improve its probability of success in mounting its defenses. These investigations cover substantive topics spanning many traditional disciplines, including genetics, nutrition, pharmacology, virology, and bacteriology, in addition to traditional medical specialties, such as internal medicine, hematology, radiology, and pathology.

The complexity of the scientific questions being asked has prompted an increase in the amount and types of data that are needed to address scientific issues. These data, however, are not unique to individual studies. In fact, most data that are collected are needed first by the clinical health-care providers to manage the patient's clinical treatment and then are used repeatedly as part of the body of data needed to evaluate research hypotheses (Flournoy et al., 1977; Flournoy and Hearne, 1982).

3. DISPERSED AND MULTIPLE USER GROUPS

The challenge to organize and manage a shared database to support a multidisciplinary research environment arises not only from the need to coordinate the many users of the database system, but also to provide support for the many viewpoints that users take of data and a database system. The quality-control considerations for specific classes of users are elaborated upon in Sections 4, 5, and 6 below. In this section, we describe the user classes. These classes are distinguished by type of need and level of responsibility for data, as shown in Table 1.

The first distinct classification is users who have short-term and immediate needs for information with which to operate the research facility. These needs contrast, and may conflict, with the need of researchers for strategic information with which to formulate and evaluate scientific hypotheses. A major consideration in developing quality-control procedures is that the researchers' vested interest in accurate and complete data, collected across time, is different from that of the operational staff, whose interest in data dissipates as soon as the data have served their immediate operational purposes. The clinical health-care provider, doctor,

Table 1. Characteristics of System Users

Class of User	Need of Data	Data Responsibility
Operational	Short term Immediate access	Little or none
Source Data Provider	Short term Immediate access	Substantial
Database Administrator	Little or none	Substantial, indirect
Researcher	Long term	Little or none

nurse, or allied health professional needs timely access to accurate information that is currently available and relevant to the clinical management of the patient. The laboratory personnel need access to current patient information that may be relevant to the execution and management of the laboratory testing procedures and preparing and delivering laboratory reports.

User views also depend on geographical or disciplinary distinctions that are significant in the technical implementation of database systems. The fact that users are dispersed does necessitate special considerations beyond those required for systems with single data sources. However, in our experience, geographical and disciplinary differences do not create distinct classes of users with respect to system development methodology.

The geographically distributed sources of information at FHCRC had no centralized authority that would dictate procedures and standards. Departments in this research facility, like those in academia, operate with a great deal of autonomy. Therefore, quality control of shared data comes only with the voluntary assumption of responsibilities by the participating disciplines. The central database administration can assume partial responsibility: it can be responsible for programming data-collection and verification routines to ensure that data entering the database are consistent with the structure of other database entries and that the data collected are plausible; it can assume responsibility for programming predefined database-retrieval routines and reports by which data can be verified at their sources on terminal screens and on paper; and it can be responsible for the integrity of the data once they are entered into the computer system. But responsibility for the accuracy and completeness of data as they are entered into the database management system, as well as the timely entry and verification of those data, rests with the participating disciplines.

The researcher needs to have access to both current and historical information from the multiple clinical and laboratory sources, together with detailed definitions of what data are available and how these data were recorded. This information is used by the researchers to design prospective comparative studies and retrospective exploratory studies, execute procedures to build datasets for an analysis, and analyze these datasets. By employing parallel and nested research procedures, understanding of the underlying disease processes is more timely and efficient.

4. THE OPERATIONAL OR CLINICAL PERSPECTIVE

To develop a sense of the interrelationships among the network of data and the research being conducted at FHCRC, let us look at the environment from the operational or clinical perspective. All patients admitted to the FHCRC have a lethal disease state and a low probability of long-term survival. When the survival probability for a patient with a particular disease gets up to .70 at 5 years after treatment, the FHCRC has historically discontinued treatment of those patients. It is assumed that such patients can be treated at cancer treatment facilities, and they no longer need the experimental research facilities at the FHCRC.

Patients who do enter the FHCRC face three major causes of death following treatment for cancer. First, the disease may not be completely eradicated, and a patient may relapse and die of the malignancy. Second, patients may contract pneumonia caused by viruses that are lethal only among people who have received heavy doses of chemotherapy and/or irradiation or people who are otherwise immunosuppressed, such as people with AIDS. Finally, organs of the body—such as the heart, liver, kidney, or brain—may fail as a direct result of the toxicity of the treatment that the patient received in an effort to kill cancer cells.

On admission to the FHCRC, patients are randomized or placed on treatment schedules from a network of experimental plans. Individual experiments may include randomization to different treatments or be strictly observational. Some plans attempt to identify treatments that reduce the risk of death from one of the three primary causes, others attempt to treat predisposing complications, and yet others attempt to describe the longitudinal behavior and interaction of cells. These experimental plans often involve more than one aspect of patient care delivery. For example, they may include procedures by which a course of drug therapy and some type of physical or radiation therapy are prescribed, a schedule for drawing and testing samples from the patient, and a schedule for tests with diagnostic devices. These plans are designed and then refined based on accrued experimental results, complications that are observed, and the evolution of new research hypotheses. The information gathered from these plans helps to illuminate the underlying disease process and improve treatment efficacy.

Patients are included in all studies for which they qualify, based on predisposing characteristics such as their current disease state; age; sex; past treatment history; evidence of bacterial, fungal, or viral infection; and nutritional status. In this way, a patient is likely to be on many different experimental plans at a given time. In fact, over the average 90-day inpatient stay at FHCRC, a patient will have participated in about 40 studies.

As an example, consider a patient who has received a randomly selected treatment for cancer. This patient will also be participating in a randomized comparative clinical trial aimed at preventing viral infection; if the patient goes on to get infected, a second randomization will determine which treatment will be used to treat the infection. This patient may also be tested repeatedly in a study of the epidemiology of a virus. At the same time, the research nutritionist may be testing blood samples to document the quantity and distribution of amino acids and essential fatty acids, trying to relate the patient's nutritional status to the body's ability to fight infection and disease relapse. And so on.

From the operational perspective, that of the health-care professionals, the shared database is the repository of the subject's presenting characteristics, treatment assignments, and clinical and laboratory data. These data are most useful when sorted in descending sequential order of chronological time and when they are readily available for consultation regarding the clinical treatment of the patient. As shown in Figure 2, the operational perspective includes data from many sources, but extends from the present back only a short distance in time.

5. THE SOURCE DATA PROVIDER PERSPECTIVE

We define the source of data to be the organization, unit, or individual that takes measurements or makes observations that are recorded in the

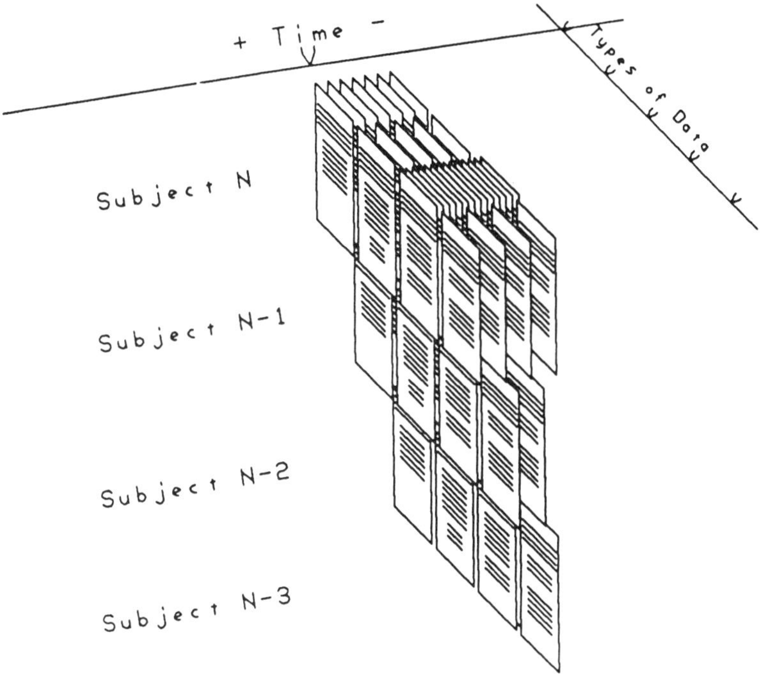

Figure 2. Operational perspective.

database. Figure 3 shows that the source data provider is responsible for a subset of the data types in the system, but needs to have access to these data for a longer time than does the operational staff. The source data provider also may have need for more detailed information than that provided to the operational system user.

As part of the basic design of the database system, we have promoted an environment in which the source data providers share responsibility for the quality of the data that they generate with the database administration and with other system users. Quality control is designed into the data-collection, management, and dissemination procedures. The maintenance of quality data cannot wait until a study terminates, because the data relevant to one study cannot be separated from the data relevant to other studies or from patient care.

At the time a data set is designed or modified, the database administrator and individual data source representatives must work together to define the meta-data, the *data about the data*. Needed meta-data include data dictionary entries in which variables are given names,

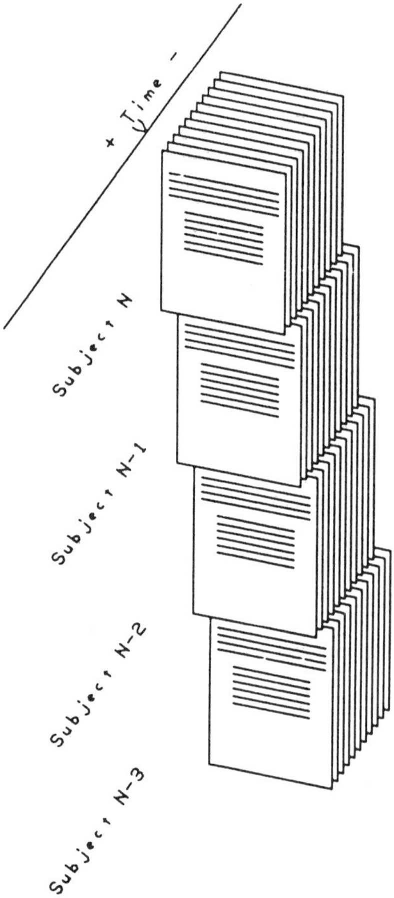

Figure 3. Source data provider perspective.

plausible field values are defined, and records are formatted. It also includes detailed verbal descriptions of each data element and the procedures, both manual and automated, that will be used to manage the data set once data accretion has started.

The joint efforts of defining the meta-data and data management procedures are, perhaps, the most critical step in establishing a system that will have quality data consistently and that can be used readily by a sizable population of scientists. The database administration works to ensure that the result of the design effort is consistent with the overall architecture of the database, that definitions are unambiguous, and that the required system implementation techniques are consistent with available hardware, software, and programming personnel. The data source representative—generally a health-care provider, senior laboratory technician, or research scientist—works to ensure that the data needed for entry are available to the personnel responsible for their capture, that data of interest from the various user perspectives are not omitted from the design, and that the designed system will have a positive impact on the organization and operation of the source data unit. This data system definition effort will involve compromise, with the understanding that there are limits to how much either the data source representative or the database administration can compromise and still have a viable operating unit.

Scheduled periodic review of individual source data systems, data sets, meta-data, and data management procedures help to ensure that this portion of the database system is being used properly and that the meta-data and management procedures still reflect the needs and operating conditions at the data source.

6. THE RESEARCH PERSPECTIVE

The research or strategic perspective on the database is in many ways the most global. Data from the many sources, crossed by the many subjects, crossed by time, define the body of evidence that is available for exploring research hypotheses. Expanding research beyond the areas currently supported by the database system maps directly to increasing the scope of source data collection. Thus, the researcher shares interest in future data, as well as in existing data, with the database administration, as shown in Figure 4.

For the scientists, source data exist on-line together with their defining meta-data. Together, they should provide all information necessary for quality use of the data. Patient demographic or core data, clinical data, and laboratory results, all of which have been verified for accuracy and completeness by other database users, are available to the research scientist. The meta-data present an unambiguous definition of the data that are available, what values can be expected, and what each data field means. The scientist who is formulating and testing hypotheses does not need full mastery of all the relevant disciplines to explore ideas and interesting lines of inquiry.

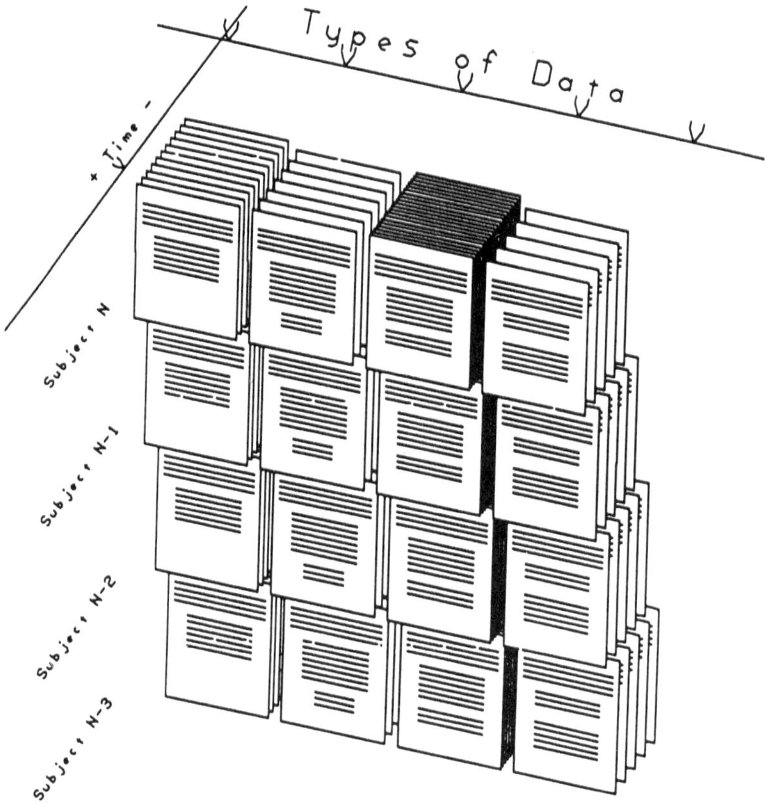

Figure 4. Researcher and database administration perspective.

Since data are captured as close to the point of generation as possible (rather than being transcribed and aggregated, introducing transcription error and loss of information), the fullest measure of accurate information possible is transferred from the data source to the database.

7. THE DATABASE ADMINISTRATOR PERSPECTIVE

One of the critical aspects of a system that captures data at the source is the ability of knowledgeable people to validate and verify its correct operation. The correctness of operation from the database administration perspective is twofold: Only data that conform to the meta-data definitions are allowed into the database and, once data are committed to the database, they are available for reporting, correction, or deletion as the need arises.

From the perspective of those who generate the source data, data management is concerned with data accuracy and data-set completeness. The meta-data definitions can restrict data to the realm of plausibility. An internal review procedure is required to verify that the correct entries were made from among the range of plausible entries. The completeness of data sets must also be established. It is only the systems users who can verify that the data in the database reflect all the data that are currently available.

It is extremely valuable to plan for evolution in the database system. Initially, a computer system designed to capture data at the source often will reflect its origins as a manual system. As the personnel at the source become comfortable using the computing system and gain understanding about how the system works, they will have suggestions for improving the utility of the system for their needs.

With time and experience, the role of the database administrator appears to shift slightly, to one of data-processing educator, as well as application design analyst. The database administrator needs time to examine the physical and operational environment of the persons doing data entry. An understanding of the user's environment provides an informed basis for, and lends credence to recommendations regarding alternative system enhancements. By encouraging and facilitating the use of the database by the personnel who are entering data at the source, the power of information as an incentive for quality control can be brought to bear and system utility enhanced.

Consider the many routine calculations that must be done in a laboratory for test results to be made relevant to patient care. One such calculation that is repeatedly required is the calculation of the number of days between the date of some treatment and the date blood is drawn for the test. Computing the number of days between two dates manually is a cumbersome and highly error-prone activity. Performing such calculations and providing reports may seem like a small thing, but the incentive this provides for timely, accurate data is dramatic. A system that improves the working conditions of the people who have responsibility for it gains their respect and appreciation and encourages their pride in a quality operation.

8. A GUIDELINE FOR QUALITY SYSTEM DEVELOPMENT

The system development methodology we recommend for designing systems that capture data of high quality is sketched below. First, identify the sources of required data elements. Data that are available at a single location at the same time are the appropriate units of data to be included in the database as a record or relation. Then, for each record, complete the following tasks:

1. Define the size and possible values for each data field within this record.
2. Define the identifier fields that will bring the data into conformance with the data access and control characteristics consistent with the database management system.
3. Define the terminal screen image for data entry.
4. Define the data-management report formats, both for displaying on terminal screens and printing on paper, to assist the source data system personnel to verify data accuracy and completeness.
5. Define the reports needed by database users external to the source data-generating unit.
6. Enter the meta-data into the software generation system, along with any special notes on data sampling, data coding, data entry, and data management. Make meta-data available to self-help facilities.
7. Generate and test the data-capture programs.
8. Generate the data management and external reporting programs.
9. Train and test personnel at the data source to use the data-capture, reporting, and self-help facilities.
10. Notify users external to the source unit of new data availability, how to access meta-data, and the names of predefined reports and retrieval routines that are available.

Several data items are helpful for maintaining quality data when added to every data record: the date and time that data are committed to the database, the identification of the person who entered the data, and the status of the data (i.e. new, changed, or deleted data records) when they were committed. This information is generated by the computer system and is recorded in hidden fields in the data record. As part of the suite of data-management reports, data can be retrieved from the database based on ranges of values for these fields. With this facility, it is possible to identify who has done what to which data when. This provides capabilities for data-set management and audit that are not available in manual systems.

9. CONCLUSIONS

The quality of data in a shared central database can be maintained at a high level of accuracy, even with a relatively autonomous user population, if the data system is designed to provide operational incentives for all classes of system users. These incentives include reduced manual records keeping, reduced manual calculations, and automated reports for the source data provider; improved organization, integration, and presentation of the most current data from multiple sources for the operational users; and ready access to well-defined, accurate, and longitudinally consistent data from multiple sources for immediate decision-making and for structured scientific experimentation. These operational incentives are the stimulus for gaining the cooperation and support needed for a shared database.

By accepting the operational incentives offered by a shared database system, source data providers are also assuming responsibility for the data they generate. The central database administration should assist the users in developing procedures, both manual and automated, that ensure that database entries are plausible. But it is the peer pressure between the data user population and the source data generators that ensures timely, accurate, and complete data in the shared database.

The function of the database administration is to negotiate a balance between the needs and desires of source data providers, database users, and data-processing constraints. This is a technical, educational, and managerial challenge for those responsible for the development, operation, and overall integrity of the shared database facility (DeMarko, 1979; Martin, 1983; Yourdon and Constantine, 1979). Quality control has been designed into the system architecture of the shared, interdisciplinary database at the FHCRC.

REFERENCES

DeMarko, T. (1979). *Structured Analysis and System Design*, Prentice-Hall, Englewood Cliffs, New Jersey.

Flournoy, N., Hearne, L. B., Schoch, H. G., and Erne, K. (1977). Data

Management in Medical Research, in *American Statistical Association: Proceedings of the Statistical Computing Section*, pp. 206–211.

Flournoy, N. and Hearne, L. B. (1982). Effects of Database Management on the Organization and Administration of Clinical Trials, in *Proceedings of the First LBL Workshop on Statistical Database Management*, H. K. T. Wong, ed., Menlo Park, California, pp. 351–352.

Martin, J. (1983). *Managing the Database Environment*, Prentice-Hall, Englewood Cliffs, New Jersey.

Yourdon, E. and Constantine, L. L. (1979). Structured Design: *Fundamentals of a Discipline of Program and System Design*, Prentice-Hall, Englewood Cliffs, New Jersey.

6

Management of Data Quality in a Long-Term Epidemiologic Study: The Framingham Heart Study

KEAVEN M. ANDERSON, DEBORAH C. DUMPHY and PETER W. F. WILSON Framingham Heart Study, National Heart, Lung, and Blood Institute, Framingham, Massachusetts

1. INTRODUCTION

The objective of the Framingham Heart Study is to find associations between suspected risk factors and the subsequent development of heart disease. Collecting data of good quality is an important part of achieving this goal. In this chapter, we describe some successes and failures of the Framingham Study in collecting high-quality data. We also discuss where the study places special emphasis on collecting good data.

The Framingham Study is a 40-year-old investigation into the causes of heart disease. Two interrelated groups, called cohorts, make up the study. The *original cohort study* sampled residents of the town of Framingham, Massachusetts, beginning in 1948. The survivors of this cohort of 5209 subjects return every two years for medical examinations. The second study began in 1971 and includes offspring of members of the original cohort. The 5135 members of this, the *offspring study*, return every four years.

The Framingham Study is a prospective study. That is, first it measures suspected risk factors from healthy subjects, then it follows its subjects to see who develops disease. The study has collected four types of data and continues to collect three of those four types. First, statisticians defined a population to sample from, and a list of that population was assembled. Since then, physicians have examined sample subjects at regular intervals. Medical records and death certificates provide information about new occurrences of heart disease. Finally, the study needs administrative data. Thus, the four data types are: a population list from which to sample, examination data, medical records, and administrative data.

2. SAMPLING

Information on the Framingham cohort sample and response comes from Gordon and Kannel (1959) and Gordon et al. (1968). In 1947, departments of the State Health Commissioner for Massachusetts, the Harvard Medical School, and the U.S. Public Health Service organized a prospective study of the epidemiology of heart disease. They chose the town of Framingham in part because of its size—10,000 people in the 30 to 59 years old age group of interest. Study designers also chose Framingham because it was near Boston and had participated in the first community study of tuberculosis beginning in 1917.

Doctors examined the first participant on September 29, 1948. Initial subjects were volunteers from the community. In the middle of 1949, the Public Health Service transferred the study to the National Heart Institute. Statisticians and staff completed a systematic sampling scheme by December of 1949 to replace recruitment through volunteers only. The population of Framingham was stratified by family size and precinct, as obtained from town records, and then ordered by address. The study selected a two-thirds systematic sample of households and invited all eligible members in selected households to participate. Of 6015 persons selected to participate, 5209 (74.3 percent) joined the study. Response rates were slightly higher for women than for men and for younger than for older age groups. There was also some variation in response by socioeconomic status. Although the study did not systematically collect information on the health status of nonrespondents, about 1 percent of the eligible population was known not to participate because of health reasons. Nonrespondents died at a slightly higher rate than did respondents during early years of the study. Comparisons of long-term mortality between responders and nonresponders did not produce conclusive results. Initially, the original volunteers who were not later chosen in the sample were analyzed separately. However, they have long since been combined with the sampled population in analyses of Framingham data. Usually, over 80 percent of those eligible for examination have returned for exams over the years.

Examinations for the Framingham offspring study began in 1971. A reference for much of the following information is Kannel et al. (1979). Most subjects were initially in the same age range—30 to 59 years—that the cohort study began with, although some subjects were as young as 12 years. Response rates were not as high for the offspring study as for the original study, probably because not all eligible people lived in the town of Framingham at the time the offspring study began.

The study consisted of three different samples. Of the 1644 spouse pairs in the original Framingham cohort, it was not known for 42 (2.6 percent) if they had any children. Of the other 1602 spouse pairs, 1404 were known to have at least one child (natural or adopted). Members of 1202 (86 percent) of these families have participated in the offspring study. However, only 2656 of 3717 (71 percent) of the children in these represented families are in the study. These 2656 subjects represent the first part of the offspring sample. The number of children who had died from coronary heart disease (CHD) before the offspring study began is not available. This group is a possible source of sampling bias in the study of CHD. The offspring study includes 899 children of 378 cohort members whose spouses were not in the original cohort study. This group was included because the parent in the study either had abnormal lipoproteins or CHD at the 10*th* biennial cohort examination. Response rates in this group are not immediately available. Finally, any spouse of an individual eligible for the offspring study was eligible to join the offspring study. This amounted to an additional 1580 subjects. Response rates in this group are also not immediately available. The study has not compared the mortality of reapondents vs. nonrespondents, as it did with the cohort study. Of the 5135 who attended the initial offspring examination, roughly 75 percent attended the second and third examinations.

Although there were some possible sampling biases, the original Framingham cohort is probably as good a sample of a free-living population as one could expect. It is not as clear what the Framingham offspring sample as a whole represents, because of the different components of the study and the less well measured response rates. Combined subsets of the two studies compose a group very useful for family studies of heart disease. The Framingham Heart Study is a study with a nonrandom sample size of one—i.e., one community. Other community- and employment-based studies have confirmed the most important results found in the Framingham Study.

3. RISK-FACTOR MEASUREMENT

Data come from several types of sources in the Framingham Study. Specific examples are: Subjects report how many cigarettes they smoke, a laboratory measures blood cholesterol, and physicians measure systolic blood pressure during a physical examination. The electrocardiogram is a machine measurement set up by a technician and read by a physician. The study has measured each of these items at most or all Framingham examination cycles since the study began. A prerequisite to obtaining these data is to get study subjects to come to clinic examinations. One problem is that as patients move from Framingham and grow older, it becomes more difficult to get them to come to examinations. This raises the question of how representative the study sample is in later examinations. One advantage the Framingham Study has in this regard is a loyalty among study subjects that has built over the years. This may result in better sample response than would a newly initiated study of the elderly.

3.1 Blood Pressure

Physicians measure resting blood pressure twice at each subject exam, and a nurse measures it once. Physicians and nurses receive no training for blood-pressure measurement specific to the Framingham Study, other than patient and cuff position at measurement. The study has standardized both equipment and number of measurements taken throughout the course of the study. Blood pressure is unique in this respect, in that the study has replicated no other measurement at each exam and has ever replicated only a few measurements.

Sources of variation in blood-pressure measurement over time include (1) between-exam trends and variation, and (2) measurement error and other within-examination variation. The standard deviation for within-examination measurement of systolic blood pressure is approximately 8.5 mmHg (Gordon et al., 1970). This can cause substantial misclassification error in labeling subjects near cutoff points for classification of hypertension. Within-examination variation for subjects is not correlated between exams. That is, if subjects have widely varying blood pressure readings at one examination, they often have consistent readings at the next examination.

Carroll et al. (1984) used an errors-in-variables model in a study of the association of systolic blood pressure with the development of coronary heart disease. This study suggests that there are important biases associated with not taking into account measurement errors. However, a practical form of analysis that does this is not available. The mean measurement at each examination would be preferable to an individual measurement because of lower error. This would reduce bias in the estimation of blood-pressure effect. In summary, a study should replicate important measures with significant measurement error when it is practical.

Blood-pressure measurements taken in different exam cycles of the study are not strictly comparable. During the first three examinations, cohort blood-pressure means decreased with time. This is called the *white coat effect*, assumed to be associated during the initial exam cycles with anticipation of a doctor's visit. It is not thought to be a true population blood-pressure change and is in contrast to increases in blood pressure with age at later examinations (Gordon and Shurtleff, 1973). Treatment of hypertension was almost nonexistent at the beginning of the study, but both treatment and control of hypertension have increased in recent years (Shea et al., 1985). These problems have often been ignored in analyses.

3.2 Cholesterol

Cholesterol measurement has changed somewhat over the course of the study. During the first examination cycle, a calorimetric method was used. Since then, the study has used the Abell-Kendall (1952) procedure. More recently, this procedure has evolved to a less labor-intensive and more machine-intensive version of this technique (McNamara and Schaeffer, 1987). The study can now standardize cholesterol measurement using a national standards laboratory (Lipid Research Clinics, 1974). Offspring subjects have fasted before each examination. The cohort study has measured cholesterol of blood from nonfasting subjects, with the exception of one exam cycle.

Hypotheses about which measurements of blood lipids should be used to assess cardiovascular risk have changed over time. During the 1950s, there was a debate over whether measuring fractions of lipids was preferable to measuring just a total cholesterol value. During that period, total cholesterol was the preferred measurement. During the 1970s, lipoprotein cholesterol fractions were again measured. Various studies since then (e.g., Gordon et al., 1977) have suggested that subfractions of high-density and low-density lipoprotein cholesterol are important.

3.3 Cigarette Smoking

Cigarette smoking is one of many items that the study relies on subjects reporting accurately. Some subjects do not report their smoking habits truthfully. Studies of disease incidence use only reports given when subjects were healthy. Thus, there is probably no association of these misclassifications with coronary heart disease.

3.4 The Electrocardiogram

Many Framingham reports have studied various measures from the resting electrocardiogram (ECG). The ECG is the primary diagnostic tool for myocardial infarction. The medical profession had not standardized the interpretation of electrocardiograms when the Framingham Study began. Thus, the study defined its own criteria for various ECG diagnoses. Rarely has there been any quality control between different readers of the ECG. An exception is the diagnosis of myocardial infarction and coronary insufficiency (unstable angina), where two of three doctors must agree on the diagnosis.

There are substantial differences between different physicians' interpretations of some ECG diagnoses. At least one study exists in which one physician read all the cardiograms in the study (Higgins et al., 1975). Measurement error weakens relations with outcome in a prospective study. If an association still exists after this weakening, it may still be important to report it even though it cannot be measured precisely (e.g., Kannel et al., 1987).

4. FOLLOW-UP DATA

Coronary heart disease (CHD) is the primary endpoint, or 'event,' of interest in the Framingham Study. CHD includes myocardial infarction, angina pectoris, coronary insufficiency (unstable angina), sudden death not attributable to causes other than CHD, and non-sudden CHD death. Dawber (1980) documents diagnostic criteria for these events.

The study finds out about coronary events from 1) regular examinations at the study clinic, 2) hospital and physician records, and 3) death certificates. We find hospital records through routine surveillance at the local hospital, or requests after a Framingham examination during which hospitalizations are reported. Also, if a subject refuses or is unable to attend an examination, the scheduler asks them about hospitalizations. Relatives of mentally incapable or recently deceased subjects inform the study of hospitalizations and illnesses. A clerk reads the Framingham newspaper's obituaries daily to find study deaths. Failing this, the study is also planning to use the National Death Index to find records for lost subjects.

A review of information sources for diagnostic information for the first 10 years of the heart study (Friedman et al., 1967) found the following. The study would not have found 46 percent of the CHD cases without subject reexamination in the study clinic. The study would have missed 7 percent without physician and hospital records. Without review of death information 13 percent of cases would have been lost.

A substantial subcontract to the study has recently found many lost subjects, especially from the offspring study. The result is that only about 3 percent of the study population remains lost to follow-up. We will not discuss further the substantial problem of tracking down 'lost' subjects.

Both classification and dating errors can occur in the recording of coronary events. Two of three physicians on a panel must agree on any positive diagnosis. The doctors on the CHD review panel vary somewhat from week to week. Over the course of the study, the pool of doctors available for review has changed completely. The criteria for diagnoses are supposedly uniform. However, some diagnostic tests available now (e.g., enzyme tests for myocardial infarction) were not available at the beginning of the study.

It is easy to date most coronary events exactly. For instance, most myocardial infarctions (MIs) cause chest pain, a hospital visit, and either an electrocardiogram or enzyme test to confirm diagnosis—all within the same day. Roughly 15 percent of myocardial infarctions produce no symptoms recognized by the victim. Physicians discover these when reading a later electrocardiogram (Kannel and Abbott, 1984). In such a case, only the dates of two electrocardiograms, one before the MI and one after, provide information on when the event occurred. The original cohort study recorded only the midpoint of these dates electronically. This dating has not resulted in missing data for many analyses of the cohort study because of the type of analysis used: The study takes ECGs every two years, and most outcome studies use only interexam interval incidence for outcomes. In studies of time to event, however, it might be more desirable to use methods that allow for interval censoring. Analysis of time to event is desirable in the offspring study, where there were eight years between the first two exams and four years between later exams. The offspring study records the closest dates known to be before and after an event with unknown date. This makes it possible to use statistical methods that eliminate biases introduced by assuming that an inexact date is exact.

5. DATA MANAGEMENT

Initially, the study recorded data on printed forms during examinations and later transferred them to a code sheet for keypunching. Clerks double-checked this transfer. In later years, staff designed printed forms on which the data were recorded at exams, so that data could be keypunched directly from them. That is, the forms became self-coding. Keypunchers have always entered data twice for verification. Before keypunching, clerks have checked forms by hand for completeness and possible inconsistencies.

In the early 1980s, the first keypunching program for data entry that contained some intelligence for making some of these consistency checks was written. Programmers later wrote an intelligent data-entry program for a personal computer that could be used for direct data entry by physicians and nurses. The study required a program to demand corrections for inconsistencies. It also had to be easy to use. Producing such a program was time-consuming and expensive. Physicians and data managers encountered many problems, and this effort was abandoned. Another study sponsored by the National Heart, Lung, and Blood Institute has successfully entered examination data directly into a computer (Christiansen et al., 1988).

Examiners now enter data on self-coding forms that contain column numbers. Clerks check these forms by hand for completeness and inconsistencies. Keypunchers enter the data twice using a word processor that constantly shows the cursor column. Physicians prefer this method, because they are used to it. Keypunchers find it faster than screen-oriented programs. Programmers no longer have to write programs for data entry. After data entry and verification, programmers read data into a statistical analysis package to make further range checks and consistency checks. If they find problems, clerks recheck paper forms for resolution.

Since September 1986, the study has had a multiuser computer with database software on site. Programmers have developed applications for two purposes: (1) data entry for simple data forms for special substudies that tend not to change between examination cycles, and (2) administrative data management. Below, we focus on some of the administrative data collection that has a very important relation to data quality for the study.

The initial administrative application developed for use by clerical personnel was to make available the study roster. This includes names, addresses, identification numbers, survival status, and morbidity follow-up status. First-order family relations between study members are available. The advantage of this system is that it makes available the most current information for any person who may need it. People who review event data, update vital status, schedule examinations, and others use these data. Previously, persons who scheduled examinations carried a list of subjects with phone numbers. When they could not reach a subject to set up an examination, there was nothing on the list to say who to contact to find the person. When a phone number or address changed, a clerk entered the change on a computerized list at some later time. The database application makes immediately available information on family members in the study, friends, and subjects' physicians. These data were available previously only in card files and subject folders. These different sources did not always have the same information. The increased access for retrieval and update of these data has improved the quality of contact information.

6. STAFFING AND TRAINING

Physicians/investigators and a statistician have worked on site at the study since its beginning. Several key staff members have spent over 20 years of their careers working primarily on the Framingham Study. Medical personnel assigned on site to the study include three physicians, two nurses, and four technicians. Other on-site personnel include a statistician, two programmers, one keypuncher, two-and-a-half receptionists/schedulers, two people responsible for review of follow-up, and two people responsible for exam data management. There are also several staff members working part-time examining subjects or supervising specific parts of the study. The National Heart, Lung, and Blood Institute has provided substantial statistical and data-management support off site throughout the history of the study.

Training is most often carried out by word of mouth. Written criteria exist (other than those given on data-collection forms) for some important data items. Observation and questioning provides most training for physician or nurse examiners. The same can be said for the use of variables in data analysis. Thus, the study has been very dependent on long-term staff members familiar with problems and strengths of the study.

7. SUMMARY

The Framingham Study has needed to collect different types of data well. The study needed a representative sample of a free-living population of both men and women in an age group with relatively low prevalence and high incidence of coronary heart disease. It needed to measure consistently a set of risk factors that were suspected at the outset of the study. The study needed to review complete, consistently measured, and documented follow-up data for coronary heart disease.

Early studies indicated that the original cohort study begun in 1948 was representative. As the cohort ages, the question of representativeness arises again, however. It is not as clear what the offspring-study sample represents. The offspring study is still a valuable resource for family studies and for studies of risk factors not measured in the original cohort. The strength of risk-factor measurement is that the study has measured a core set of variables reasonably regularly and consistently over time. Difficulties in this area include some changes in measurement and medical practice over the years. Another problem is that there is a temptation to ask every question and make every test under the sun. The quality of data for the wide variety of items collected is far from uniformly good. This is largely because of a lack of resources to review each measurement method well prior to its use. The biggest danger associated with this problem is studying an important risk factor and measuring it so poorly that no association with disease is found. Finally, the problem of adequate follow-up for both morbidity and mortality has been well understood and addressed in this study.

REFERENCES

Abell, L. L., Levy, B. B., Brodie, B. B., and Kendall, F. E. (1952). A Simplified Method for the Estimation of Total Cholesterol in Serum and Demonstration of its Specificity. *J Biol Chem*, 195:357–366.

Carroll, R. J., Spiegelman, C. H., Gordon, K. K., Bailey, K. T., and Abbott, R. D. (1984). On Errors-in-Variables for Binary Regression Models. *Biometrika*, 71:19–25

Christiansen, D. H., Hosking, J. D., and Dannenberg, A. L. (1988). *Computer-Assisted Data Collection in Multicenter Epidemiologic Research: The Atherosclerosis Risk in Communities Study*. Submitted for publication.

Dawber, T. R. (1980). *The Framingham Study. The Epidemiology of Atherosclerotic Disease*. Harvard University Press, Cambridge, Massachusetts.

Friedman, G. D., Kannel, W. B., Dawber, T. R., and McNamara, P. M. (1967). An Evaluation of Follow-up Methods in the Framingham Heart Study. *Amer J of Public Health*, 57:1015–1024.

Gordon, T., Moore, F. E., Shurtleff, D., and Dawber, T. R. (1959). Some Methodologic Problems in the Long-Term Study of Cardiovascular Disease: Observations on the Framingham Study. *J of Chronic Diseases* 10:186–206.

Gordon, T. and Kannel, W. B. (1968). *The Framingham Study. An Epidemiologic Investigation of Cardiovascular Disease. Section 1† Introduction and General Background*. U.S. Department of Health, Education, and Welfare.

Gordon, T., Sorlie, P., and Kannel, W. B. (1970). Problems in the Assessment of Blood Pressure: The Framingham Study. *International J of Epidemiology*, 5:327–334.

Gordon, T. and Shurtleff, D. (1973). *The Framingham Study. An Epidemiologic Investigation of Cardiovascular Disease. Section 29. Means at each examination and inter-examination variation of specified characteristics: Framingham Study, exam 1 to exam 10*. U.S. Department of Health, Education, and Welfare, Publication [NIH]74–478.

Gordon, T., Castelli, W. P., Hjortland, M. C., et al. (1977). High-Density Lipoprotein as a Protective Factor Against Coronary Heart Disease: The Framingham Study. *Amer J Med*, 62:707–14.

Higgins, I. T. T., Kannel, W. B., and Dawber, T. R. (1965). The Electrocardiogram in Epidemiological Studies. *Brit J of Preventive and Social Medicine*, 19:53–68.

Kannel, W. B. and Abbott, R. D. (1984). Incidence and Prognosis of Unrecognized Myocardial Infarction. An Update on the Framingham Study. *New England J of Med*, 311:1144–1147.

Kannel, W. B., Feinleib, M., McNamara, P. M., Garrison, R. J., and Castelli, W. P. (1979). An Investigation of Coronary Heart Disease in

Families. The Framingham Offspring Study. *Amer J of Epidemiology*, 110:281–290.

Kannel, W. B., Anderson, K. M., McGee, D. L., Degatano, L. S., and Stampfer, M. J. (1987). Nonspecific Electrocardiographic Abnormality as a Predictor of Coronary Heart Disease: The Framingham Study. *Amer Heart J*, 113:370–376.

Lipid Research Clinics Program, Manual of Laboratory Operations, Vol. 1, U.S. Dept Health Education and Welfare Publ. No. [NIH] 75–628 (1974).

McNamara, R. R. and Schaeffer, E. J. (1987). Automated Enzymatic Standardized Lipid Analyses for Plasma and Lipoprotein Fractions. *Clinica Acta*, 166:1–8.

Shea, S., Cook, E. F., Kannell, W. B., and Goldman, L. (1985). Treatment of Hypertension and its Effect on Cardiovascular Risk Factors: Data from the Framingham Heart Study. *Circulation*, 71:22–30.

7
Issues Related to Field Reliability and Warranty Data

JEFFREY A. ROBINSON and GARY C. McDONALD
General Motors Research Laboratories, Warren, Michigan

1. INTRODUCTION

A major challenge confronting automotive manufacturers is, quite simply, measuring the quality of their products. How does the vehicle meet customers' expectations upon delivery—and after months and years of service? Many organizations are extensively involved in the collection, tabulation, and reporting of such information. In this chapter, we will discuss various types of data sources and indicate strengths and limitations that, inevitably, qualify the application of the results in practice. While our discussion will cover survey data and field data, the focus will be on warranty data, which has had varied applications and treatment in the statistical literature.

Warranties have become a critically important segment of the automobile business environment. Figure 1 displays qualitatively the business facets that are impacted directly by the specification of warranties in the automotive industry. These facets are not, of course, totally separate and disjoint as depicted in the figure. In reality, these various aspects are interrelated and closely linked. However, viewing these facets as distinct will enable us to place into context much of the mathematical modeling that has been related to warranty analysis and to delineate where our particular contribution falls.

Figure 1. Business facets impacted by warranties.

Perhaps most of the published literature on warranty analysis revolves around *cost* and exploits renewal theory to estimate expected number of replacement items. Blischke and Scheuer (1975, 1981) follow this format in addition to calculating the cost of warranty policies as a function of lifetime distributions applicable to the product. Balachandran et al. (1981) use a Markov structure to estimate and analyze repair and replacement costs as a step in specifying a product warranty period. In a related paper, Maschmeyer and Balachandran (1980) propose a stochastic model that both describes the breakdown phenomena that affect products and facilitates estimating repair costs during a maintenance agreement period. In this context, maintenance agreements typically take effect when products have survived their warranty periods and then fail. Nguyen and Murthy (1984a, 1984b) construct a general model for estimating warranty costs for repairable products sold with a specified warranty period. They estimate the expected total warranty cost and prediction interval for a fixed lot size of sales and the expected number of units returned for repair. Their results are applicable to any failure time distribution and for various types of repair. Amernic (1979), in an article illustrating how matrix algebra can be usefully applied to the solution of accounting problems, uses Markov models to address the warranty-liability estimation problem.

Balcer and Sahin (1986) consider a product sold under a pro rata policy and under a free-replacement policy. They characterize all the moments of the total replacement cost in finite time under both of these policies. As noted, much of the warranty analysis is linked to renewal theory. Frees (1986) has examined alternative methods for renewal function estimation—both parametric and nonparametric. He provides a Monte Carlo study to suggest guidelines for choosing an estimator under various circumstances.

There seems to be little published on the *advertising* and *product attribute* aspects of warranties. Darden and Rao (1979) construct a linear covariate model for warranty attitudes and behavior. Gottfried and Madison (1975) examine the pattern of warranty competition and find it to be consistent from industry to industry. Simply put, one manufacturer introduces warranties or breaks the established precedent by extending the warranty period, and competitors find it necessary to follow suit. Recent warranty changes in the automotive industry seem to follow a similar pattern. While little may be published on these important facets of warranties, consumers are well aware of the advertising value of increased warranty coverage, such as in the automotive sector. Consumer surveys assessing reasons for a particular automotive purchase often cite warranty coverage as one such reason.

Dealer relations and *customer satisfaction* are two important facets that seem to get little coverage in the literature. Miller (1973) describes the features of the American Motors' Buyer Protection Plan and its broad impact on dealers and customers. He discusses associated methods to provide product problem information feedback (a facet of *reliability*), summarizes results from dealer surveys, and describes a product problem and action status reporting procedure. The latter procedure was developed to provide a uniform, coordinated method for investigating, recording, and resolving field quality problems.

Reliability, like cost, has received substantial attention in the literature. This is because, to a large extent, the related issues are amenable to a mathematical formulation. Courville and Hausman (1979) examine the warranty scope and reliability under imperfect information and alternative market structures. Yun and Kalivoda (1977) develop a simple model for estimating realistic warranty return levels corrected for human errors in the determination of product failures and the various constraints imposed by operational procedures in warranty administration. An experimental determination of probabilistic information in analyzing warranty return products is suggested, as well as a method of setting upper and lower limits on the probabilities. Moss (1978) analyzes field warranty data on the electrical components used in typical Hewlett-Packard instruments and shows that hardly any exhibit a constant hazard rate during the 12-month warranty period. We've observed similar patterns with warranty data on many automotive components. Moss suggests a Weibull model to represent such populations. Mann (1970, 1976) examines warranty periods for production lots based on fatigue-test data, including ordered sample observations from a Weibull population. Schneider and Tibbetts (1983) assess engine product reliability from warranty data. Bernath (1984) examines some principles of component and equipment screening to improve reliability and reduce warranty costs. An objective of this work was to provide guidelines to those designing screening programs within a framework of economic constraints and desired reliability. Suzuki (1985a) developed a nonparametric estimator of the reliability function under a sampling structure that can be applied to automobile warranty data. The data are assumed to consist of miles to the first repair for all repairs that occur within the warranty period (measured in calendar time) and mileage at the end of the warranty period for a follow-up random sample of cars that did not have repairs. A parametric approach is taken in Suzuki (1985b), using the Weibull and exponential models as examples.

Product *pricing* and associated warranty coverage can be closely interrelated when warranty expense is significant compared to the unit profit. Anderson (1977) describes a general decision model for the choice of a warranty period and a product price model that maximizes profit when the product has a random service life. The model includes a product-demand function that is dependent on unit price and warranty period, a warranty-rebate function, and the product-failure function. Glickman and Berger (1976) also analyze a profit-maximizing model by choosing the optimal price and warranty period. They provide sensitivity analyses for a gamma-type repair distribution. Matthews and Moore (1987) use an economic optimization approach within a monopoly context to address the problem of designing and pricing a product line of goods distinguished by different quality and warranty levels. They use new techniques for dealing with incentive constraints between nonadjacent consumer types and are able to characterize optimal allocations. Singpurwalla (1987) details a strategy for choosing an optimal warranty based on the principle of maximization of expected utilities involving both profit and costs. His approach incorporates a bivariate probability model involving time and usage as censoring variables. In the case of automotive warranties, the two variables consist of the age of the vehicle since delivery to the customer and the odometer mileage.

Menke (1969) presents a method to calculate the warranty reserve fund needed to cover the expected risks of warranty claims of nonrepairable products under pro rata and lump-sum rebate plans. He provides a parametric formulation allowing the manufacturer to manipulate marketing strategy and cost policy within the model. Patankar and Worm (1981) recognize that cash flows associated with a warranty policy are a random variable and develop prediction intervals for warranty reserves and cash flow. In their context, the cash flow is the expected total rebate to be paid during the year when customers exercise their warranties.

In this chapter, we approach warranty modeling from the pricing and accounting perspectives somewhat in the spirit of the work described in the previous paragraph. We also incorporate a component attributable to the reliability facet. Section 2 places the warranty database into context with other field data sources applicable in the automotive business. Section 3 describes the engineering, financial, and cost motivations for examining warranty data in several different ways. Section 4 deals with a new approach to forecasting warranty expense on a per-car basis. This approach involves a new structuring of the databases to facilitate efficient parameter estimation in an empirically derived statistical model. The model lends itself to a Bayesian approach, which is particularly well suited for vehicle models that are carryovers from the previous year(s). Section 5 provides a summary and conclusions of our work.

2. FIELD DATA SOURCES

Car manufacturers rely on many sources of performance information on their products in customer usage. These sources provide information about varied aspects of the business. This section contains a brief summary of some of the data sources encountered by the authors. The quality of the data for their intended purpose is an important consideration. It is probably impossible to quantify cost-to-quality ratios for these data sources, but users undoubtedly make subjective decisions about data quality in choosing data sources.

2.1 Surveys

One primary source of information comes from surveys of randomly selected customers. Surveys fall into three basic types: (1) those performed or commissioned by the manufacturer exclusively for its use, (2) those purchased by the manufacturer and shared by other manufacturers, and (3) those aimed directly at the public. The details of the structure and purpose of the first type are proprietary. The National Car Quality Tracking Study by Maritz Market Research Co. and the J. D. Power survey are examples of the second type. Consumer surveys, such as those reported in the annual auto issue of *Consumer Reports*, fall into the third type. The survey respondents in this latter instance are subscribers to the magazine and, hence, introduce a potential source of bias.

Most surveys are sent to owners of new cars. There are several different slants that a survey can take. One type generates buyer profiles, including demographics, income, and type and age of the trade-in vehicle. A survey usually addresses customer satisfaction with the product and the ownership experience, including buying and servicing the vehicle. Some specialized versions of these surveys focus on the dealer and are used to evaluate dealer performance.

There are several advantages of surveys as a data source. They are inexpensive per unit surveyed compared to other sources of information. Large samples become practical, resulting in valid statistical comparisons, even among different models, engine types, and dealers. Another advantage is that surveys can easily include information on competitive models. Samples of registered owners of new vehicles can be obtained for this purpose.

Bias due to nonresponse reduces the quality of the data. Survey firms try to minimize this problem by including small cash payments with the survey form and by recontacting nonresponders. A lack of specifics, partially caused by imprecise terminology, is another problem with surveys. Avoiding all jargon and potentially ambiguous phrasing can help, but this problem can never be eliminated entirely. For example, a customer might respond unfavorably to the "water leaks" question on a survey. This, naturally, would lead the engineer to door and window moldings, windshield fits, and similar items. A more detailed interview with the customer could reveal that the water originated as condensation from the air conditioning in the passenger compartment. This is still a problem, but not the one we might naturally consider first.

When survey information is insufficient as a basis for action, some follow-up is required. This is true for assessing customer requirements as well as for problem identification. The follow-up could be in the form of a phone interview, a letter, or a personal meeting. Follow-up occurs when there is a perception that it is the best way to get the required information. We are aware of an example where survey follow-up was instrumental in deciding to add a particular type of glove box to a new car model. The cost of survey follow-up is always a consideration, but is increasingly becoming subordinate to the notion that the only way to hear the voice of the customer is to talk to the customer. The risk of overburdening the respondent, however, is also a consideration. In the absence of follow-up information, surveys can be used to indicate general trends, which, when combined with "hard" engineering tests, help in problem resolution.

Though used to a lesser extent, surveys of owners of older cars are sometimes performed. Some collect general information about the whole vehicle, while others specialize on a particular vehicle system. The emphasis is usually on durability, with information collected on cars as old as four or five years. Owners are asked what problems they have experienced in the last six months (other surveys use one year). A given vehicle is not tracked for its entire life. Each sample is stratified to select some one-year-old cars, some two-year-old cars, etc. The whole-car durability surveys, like those for new cars, have limited value for problem identification and resolution because they lack specifics. There are, however, examples of very successful surveys for studying durability of a particular vehicle system. These surveys generate information on a large customer base cheaply and, as one user has said, they "let the customer be the judge of quality."

All long-term durability surveys rely heavily on the customer's memory. This can lead to granularity in the data. For example, when mileage at repair is asked for, relatively few report 33,000 miles, but relatively many report 30,000 miles or 40,000 miles. The main advantages of long-term surveys is that they provide information beyond the covered warranty period and on competitive vehicles, as well as a manufacturer's own.

2.2 Proving-Ground Tests

Proving-ground tests are actual driving tests done by professional drivers over designed courses and driving schedules. They are used primarily to validate designs and manufacturing processes during the product development process. However, some production vehicles are run on these tests. Any failures are reported and analyzed in detail. Sample sizes are usually small—ten or fewer vehicles of a particular model. Therefore, car-to-car variation cannot be assessed accurately. These tests are more expensive per unit than are the other types of studies. The information generated is not viewed as customer experience, but rather as a type of accelerated test information that forms benchmark criteria to which new products, similarly tested, can be compared.

2.3 Warranty Data

In terms of sheer volume, warranty provides more data than any of the other sources of information. Auto warranties have become more inclusive—but, at the same time, more complicated. More data are being stored and processed over more extensive coverage periods. (Some manufacturers have recently extended bumper-to-bumper coverage to three years of customer ownership and power-train coverage to seven years of ownership.) However, warranty systems are maintained primarily to pay dealers—not to provide feedback to engineers. The use of warranty data systems in a context for which they were not originally designed is a widely recognized problem and not easily solved. Better reporting can help alleviate the problem. The age-specific warranty analysis discussed in the next section is an example of improved reporting.

The warranty claim record documents the labor operation performed. It might not identify the failed part, the customer complaint, or the condition of parts that are replaced. For example, a labor operation to remove and replace an electronic control module (microcomputer) might reveal very little about the nature of a failure if, indeed, the module has failed. Warranty data are voluminous and not very specific; however, using them for engineering purposes is essentially free. The data exist as a by-product of paying dealers and keeping the necessary financial records.

There are some enhanced warranty data systems encountered by the authors that attempt to overcome some of these deficiencies. One such system collects additional information on all warranty claims coming into specified dealerships. Dealers are compensated for the increased labor required. The additional information includes failed parts, customer complaint, and cause of the problem. This type of system can provide useful information if expert mechanics input or oversee the input of the data. Experts are expensive, however, and if they are replaced by nonexperts, there is a danger of generating data no more useful than conventional warranty data.

Another type of system, somewhat related to warranty, tracks problems and repairs on a few hundred customer cars. Prior to selection, the customers are identified as high-mileage drivers. This produces high-mileage information relatively quickly. The tracking extends beyond the regular warranty coverage, and customers are compensated for their time and trouble by getting free maintenance and service. This type of system generates more specific data than do long-term durability surveys. It does not, however, include competitors' vehicles.

Ultimately, all field data are used to improve the products offered to the customer. But there is some overlap in content among the various sources of information. In these competitive times, the challenge is to eliminate unnecessary costs while improving the quality of the information.

3. WARRANTY REPORTS

3.1 Financial Departments (Cost)

Financial departments have some interest in tracking total dollars spent on warranty and documenting the expenditure by characteristics; e.g., by car line, by assembly plant, etc. Such reports are fairly straightforward. Most often, there is a need to express the dollars *per car* eligible to generate warranty expense.

The exposure base for such calculations can be troublesome. The number of cars produced is the easiest to use because the data are readily available. However, this measure can result in misleading conclusions when one is comparing two model years or groups that have greatly different levels of dealer inventory. The number of cars sold is slightly more troublesome to obtain, but using this measure eliminates difficulties caused by varying inventories. Cumulative warranty expense per eligible vehicle is often displayed as in Figure 2, showing cost per car at various points in time for different model years. This chart is a basis for comparison of models and leads to the natural question of forecasting lifetime costs for the most recent model.

3.2 Engineering

Engineering groups are more concerned with frequency of repairs or number of dealer visits than with the cost of warranty repairs. For warranty information to be meaningful to product engineers, it should be categorized by major systems and subcategorized into subsystems and components. Ideally, the systems should reflect the functional use and also the engineering organization. This breakout facilitates comparison with other engineering tests, such as those conducted in the laboratory and at the proving ground. This format is also used in planning reliability apportionment for future products. The apportionment of reliability to the new product begins with a statement of where the current product stands

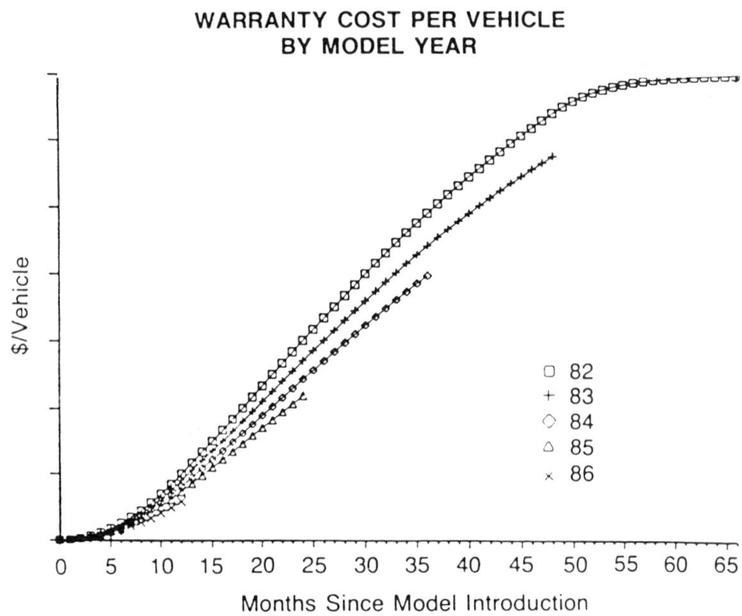

Figure 2. Cumulative cost per car by months since model introduction.

by system and subsystem. Pareto charts and top frequency lists within each major system are monitored to track problems.

An engineering analysis of warranty data requires more of a reliability approach than does a financial analysis. A type of report that we describe as *age-specific warranty frequency* is illustrated in Table 1. This technique uses the sales date and the repair order date to age-classify warranty repairs. It also uses the current date and the sales date to determine the age of each vehicle sold. It is important to note that this is done for all vehicles—not just those that have generated a warranty claim. The number of vehicles eligible in the table for each age category forms the denominator for calculating incremental warranty repair rates. This report is a grouped data variation on a type of analysis reported by Nelson (1988). The zero category represents warranty work done by the dealer

Table 1. Age–Specific Warranty Frequency Analysis

Age (days)	Number In	Number of Cars Eligible*	Number of Cases	Incremental Frequency (per 1000 cars)	Cumulative Frequency (per 1000 cars)
0	—	5000	375	75	75
01–30	1000	4500	900	200	275
31–60	1000	3500	525	150	425
61–90	1000	2500	313	125	550
91–120	1000	1500	165	110	660
121–150	1000	500	50	100	760
151–180	0	0	0	—	—
181–210	0	0	0	—	—
⋮	⋮	⋮	⋮	⋮	⋮

**Eligible* equals one-half the number in the given age category plus all those in a greater category.

before delivery of the vehicle to the customer. The incremental repair rates are almost never constant with age. They decrease with age, indicating that new vehicles generate more warranty repairs than do old vehicles. This type of report eliminates distortions caused by dealers' inventories, since it uses only sold vehicles. Similarly, varying sales patterns are handled, since the method accounts for the length of time since each vehicle has been sold. Finally, since age groups are segregated in the analysis, nonconstant repair rates are comprehended. Some other factors, such as seasonal effects, are not taken into account.

In practice, the incremental rates are accumulated, and the resulting cumulative warranty frequency is plotted against months in service. On logarithmic scales, the plots are usually nearly linear, with breaks occurring where warranty coverages change. Figure 3 is an illustrative example. It is very tempting to use simple extrapolations based on these plots to predict the cumulative number of repairs at ages greater than those that have been observed.

Despite all the advantages of age-specific warranty analysis, we have observed some difficulties putting it into practice. The method is sensitive to how the data are processed. We have observed a tendency for the cumulative frequencies displayed in Figure 3 to increase with calendar time. To illustrate: In January, we might have a warranty figure representing 90 days in customer service. In February, we update the figure. Since more cars have passed into and through the 90-day period, the figure changes—either up or down. Our experience has indicated that it almost always goes up. Figure 4 (arithmetic scales) illustrates how such curves can grow. This bias essentially destroys any forecasting capability of the method. One explanation, of course, is that the cars are actually getting worse. In general, this turns out not to be true, and the same growth occurs even when we consider only cars built in the same period.

Two other factors seem to be most important in explaining the bias. One has to do with denominators and the other with numerators. The original version of age-specific warranty analysis encountered by the authors did not use the denominator in Table 1, which counts only half the cars in a given age interval plus all the cars that have passed through the interval. The original version counted all the cars in the interval plus those that had passed through. (This version was used to generate Figure 4.) Intuitively, this overcounts the denominator, because a car in the interval for only one day has only 1/30th the chance of generating a warranty repair as a car that has passed entirely through the 30-day interval, and yet counts the same in the denominator. The overcounting is temporary since,

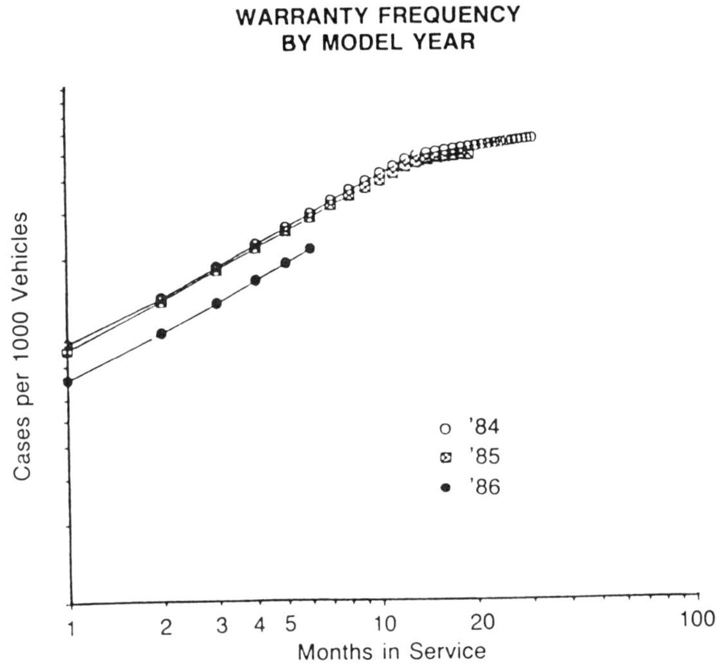

Figure 3. Cumulative warranty frequency by months in service.

eventually, all cars pass through any given age category. Since the denominators are temporarily overcounted, the repair rates are temporarily understated; hence, the growth.

This problem was soon detected and corrected. The denominators in Table 1 count only one-half the cars in the interval. This assumes that, on average, the number of days spent in a given 30-day age interval is 15. This would be true if sales were exactly uniform over time. This enhancement can be carried one step further by calculating exactly how many days each car has spent in each age interval. It turns out not to make much difference except for the last interval. This version of the denominator enhancement

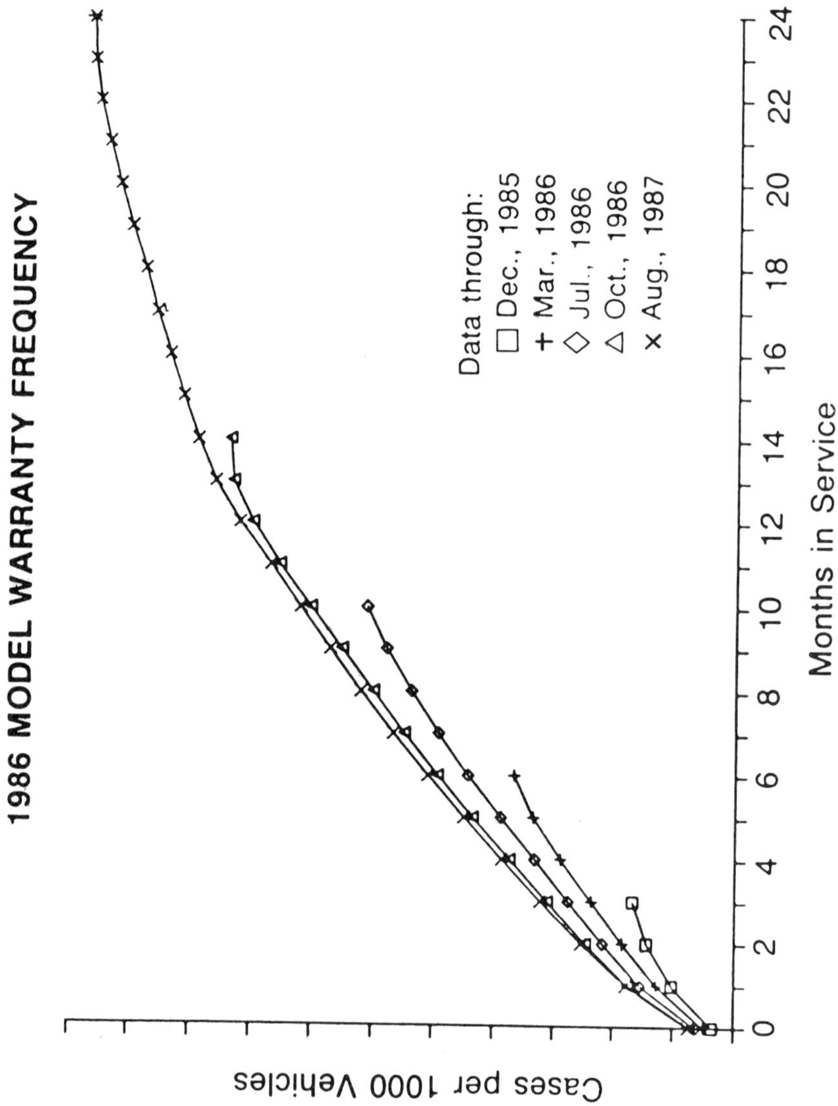

Figure 4. Example of apparently "growing" warranty rates.

was implemented on the system and seems to have reduced the growth by about one-half that experienced earlier.

Another reason for this bias has been identified that is not as easy to correct. The system that generates the age-specific warranty report is updated monthly. It allows approximately six weeks for a warranty claim to clear through the system. So, for example, an update taking place on June 15 would look for cars sold before April 30 and repair orders dated before April 30. Some claims take longer than six weeks to clear. A dealership can hold a claim, though it is not in its best interest to do so. A claim can be rejected (e.g., out-of-warranty coverage), corrected, resubmitted, and finally paid and posted on the system. For these reasons, the claims (numerators) in the updates tend to be undercounted. Again, the undercounting is temporary, since all claims are eventually settled. This contributes to the upward movement of the reported warranty repair rates.

Obviously, this effect can be reduced by either reducing the time required to clear claims or else allowing more time for them to clear. Electronic reporting systems will take care of the former in time. The latter is an interesting trade-off between timeliness of information and quality of information. We could, for example, allow an additional month for claims to clear. This would reduce the bias mentioned above, but would also reduce the amount of (apparent) information contained in the report, since moving the cutoff date from April 30 to March 31 would drop out all cars sold in April and effectively make all the other cars one month younger. More importantly, this change would delay the first report by one month. We have found it difficult to sell this type of change. This particular one is currently under study in our own organization.

Engineers in manufacturing operations have requirements similar to those in product engineering, but there are some differences. Manufacturing engineers are less interested in breakouts by major vehicle system than in breakouts by assembly plant for assembly-related items. The age-specific warranty analysis is also used in manufacturing, usually with one addition. The results are segregated by month of vehicle production. The reports plot warranty frequency at preselected times in service as a function of month of production. Common choices for times in service are 30, 60, 90, 180, and 360 days. Figure 5 illustrates the technique. This type of report is monitored regularly and looked at especially carefully after any changes are made.

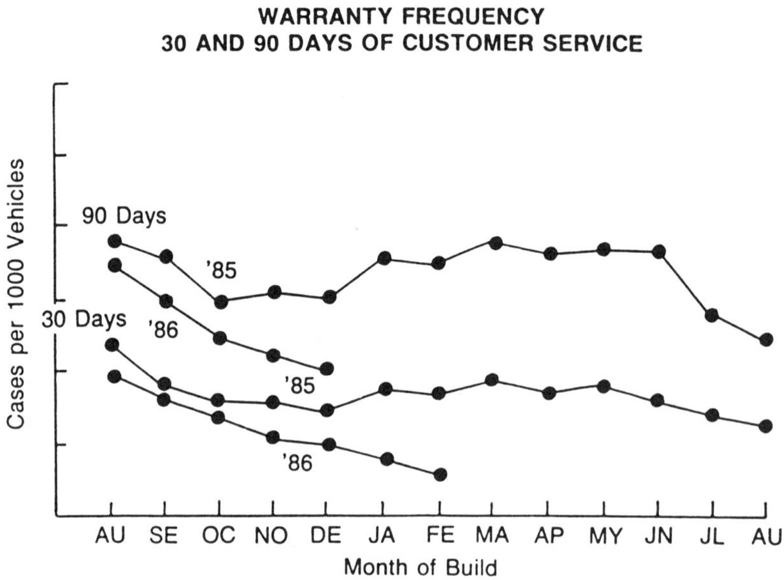

Figure 5. Warranty frequency by month of build.

4. WARRANTY FORECASTING

In the authors' experience, the demand by both engineers and accountants for warranty forecasting has increased dramatically in the last few years. In Section 3, we mentioned briefly how the age-specific warranty analysis method could be used to make simple warranty forecasts. An ideal warranty cost forecasting and reporting scheme would function as displayed in Figure 6. The initial accrual level is established using historical data on similar products and expert opinion. As actual product warranty data become available, forecasts are established together with estimates of uncertainty. As more data become available, the forecast is updated and the prediction limits narrow. If, at any time, we find the accrual level outside the prediction limits, we make a business decision to adjust the accrual level. This affects the profit statement, since the

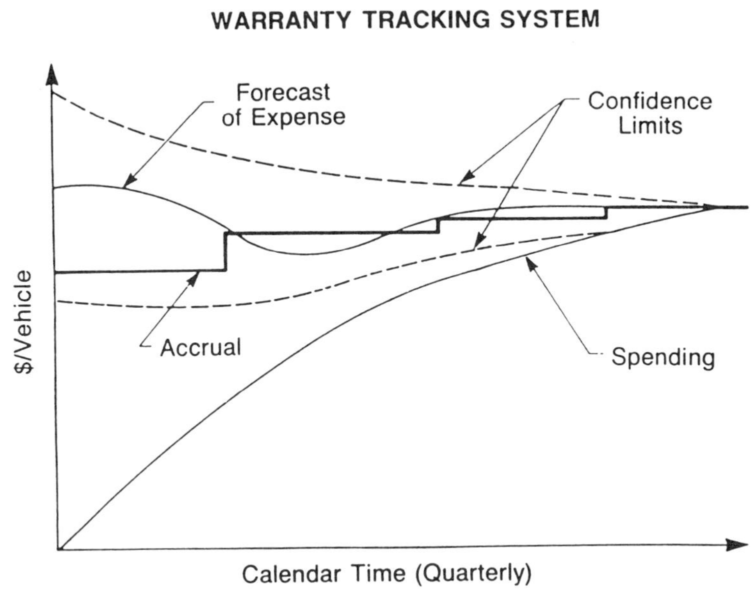

Figure 6. Ideal warranty reporting scheme.

adjustment must be made retroactively to units already sold. Eventually, spending, forecasts, accrual levels, and prediction limits converge.

We have also seen attempts to develop more sophisticated warranty cost forecasting systems. The quantity to be forecast can be represented as a sum of costs associated with cars in various age intervals over various time updates divided by the total number of cars eligible. A more detailed discussion can be found in Robinson and McDonald (1987). The individual items in the sum can be included in a regression model with predictors that include age, calendar time, and model year. One can also consider a Bayesian version that places prior distributions on the regression parameters. This is especially helpful in practice for the model year parameter, since it allows for input of test experience or expert opinion of the form, "this year will be 10 percent less than last year."

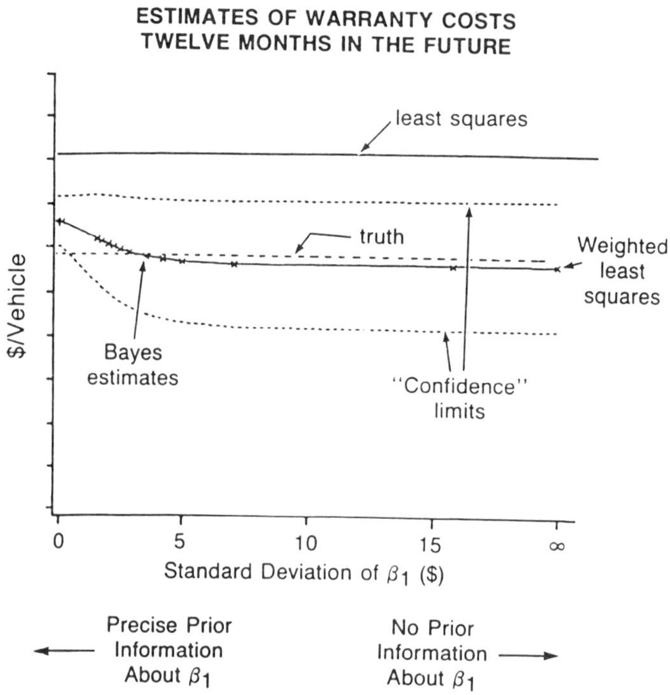

Figure 7. Bayesian warranty forecasts compared to least squares.

Figure 7 illustrates the Bayesian approach on a real data example. (We are indebted to Diane Gibbons, Lonnie Vance, Jim Wendelberger, and Mike Wincek for this figure and acknowledge their work in this modeling effort.) The problem was to forecast 12 months ahead. The true value was determined by a retrospective examination of the data. All regression parameters were given noninformative priors, except the model year parameter, which called for a moderate improvement. Figure 7 is a plot of the Bayesian cost estimate as a function of the prior standard deviation for the model year parameter. A very large standard deviation represents no prior information and, in this case, makes the Bayesian estimator essentially equivalent to the weighted-least-squares (WLS) estimator. In this case, the prior information improves the forecast up to a point, moving the forecast nearer the true value than does the WLS forecast. But as the prior variance is reduced even more, the forecast is dominated by expert opinion and begins to degrade. All the Bayesian forecasts in this case, no matter what the prior, are superior to the ordinary-least-squares (OLS) forecast.

These methods need to be tested more on real data before they can be implemented for routine use. One problem we have encountered is that the database required by the model is more extensive than any currently in place.

5. SUMMARY AND CONCLUSIONS

The published literature on warranty analysis can be interpreted in terms of the attributes and uses of warranty information. These include cost, advertising value, warranty as an attribute of the product, dealer relations, customer satisfaction, and reliability. Warranty information is one source of information for auto manufacturers on how the product performs in the hands of the customer. Surveys, both of new car owners and long-term, are another source. Surveys fall into three basic types: (1) those performed or commissioned by the manufacturer exclusively for its use, (2) those purchased by the manufacturer and shared by other manufacturers, and (3) those aimed directly at the public.

Different departments have developed their own methods for reporting warranty information. Financial departments are interested primarily in costs. They often look at cumulative warranty cost per car sold or per car produced. Engineers are more interested in the number of warranty cases. Product engineers subdivide warranty into major vehicle systems subsystems, and components. They also prefer an age-based exposure measure over other methods. Age is calculated in terms of time since delivery of the vehicle to the customer. Manufacturing engineers like to segregate the data by assembly plant and month of production within an assembly plant.

The different warranty methods of different departments are seldom reconciled, but there are exceptions. We know of one case where accountants and engineers were sending different signals to top management about a particular car line. The accountants were saying that warranty costs were going up, and the engineers were saying that warranty frequencies were going down. Both were right: But cost per warranty claim was going up fast enough to make up for the decrease in warranty frequency. This led to an investigation into why cost per claim was increasing. The usual increase for inflation in parts and labor rates did not explain it. The mix of claims had changed—more expensive claims and less inexpensive claims. This trend is related to complexity of the vehicle. After this incident, cost per warranty claim came under much closer scrutiny.

Standard accounting practice calls for warranty cost forecasts to account for future liabilities. Accurate forecasts promote good financial planning and accurate financial statements. An ideal warranty cost forecasting scheme would track spending to date, forecasts of lifetime spending, prediction limits for the forecast, and the warranty cost accrual level used in the financial statement.

Some work has been done lately at General Motors toward forecasting future warranty costs. The mathematical model used as a dependent variable the cost incurred for cars of a given model year of a given age category, in a given period of time. The independent variables in the regression model included model year, age, time, seasonal factors, and various quadratic terms. Weighted-least-squares and Bayesian methods have been used to estimate the parameters in the model. The warranty cost forecasting solution in the context of this model can be stated as a prediction interval for a total of future costs. The statistical model was formulated using actual data on a single car line. It is currently being tested on data from a wide range of model years and car lines.

A gap exists between the practice of warranty analysis and methods that have appeared in the published literature. This chapter outlines some of the practices and requirements of an auto manufacturer and describes some work under way in the area of warranty cost forecasting. Our intention is to narrow the gap somewhat and to contribute to a discussion that will direct efforts toward narrowing the gap further.

REFERENCES

Amernic, J. (1979). A Matrix Approach to Asset and Liability Valuation, *Cost and Management*, 53:25–31.

Anderson, E. E. (1977). Product Price and Warranty Terms: An Optimization Model, *Oper. Research Quarterly* (*GB*), 28:739–741.

Balachandran, K. R., Maschmeyer, R. A., and Livingstone, J. L. (1981). Product Warranty Period: A Markovian Approach to Estimation and Analysis of Repair and Replacement Costs, *Accounting Review*, 56:115–124.

Balcer, Y. and Sahin, I. (1986). Replacement Costs under Warranty: Cost Moments and Time Variability, *Operations Research*, 34:554–559.

Bernath, J. C., Jr. (1984). Reliability Improvement Through Equipment Screening, 1984 *Conference Proceedings of the Third Annual International Phoenix Conference on Computers and Communications*, pp. 208–213.

Blischke, W. R. and Scheuer, E. M. (1975). Calculation of the Cost of Warranty Policies as a Function of Estimated Life Distributions, *Naval Research Logistics Quarterly*, 22:681–696.

Blinschke, W. R. and Scheuer, E. M. (1981). Applications of Renewal Theory in Analysis of the Free-Replacement Warranty, *Naval Research Logistics Quarterly*, 28:193–205.

Courville, L. and Hausman, W. H. (1979). Warranty Scope and Reliability under Imperfect Information and Alternative Market Structures, *J of Business*, 52:361–378.

Darden, W. R. and Rao, C. P. (1979). A Linear Covariate Model of Warranty Attitudes and Behaviors, *J of Marketing Research*, 16:466–477.

Frees, E. W. (1986). Warranty Analysis and Renewal Function Estimation, *Naval Research Logistics Quarterly*, 33:361–372.

Glickman, T. A. and Berger, P. D. (1976). Optimal Price Protection Period Decisions for a Product Under Warranty, *Management Science*, 22:1381–1390.

Gottfried, P. and Madison, R. L. (1975). Warranty Analysis: Industrial and Commercial Product Reliability, *Proceedings of the 6th Annual Product Liability Prevention Conference*, pp. 37–42.

Mann, N. R. (1970). Warranty Periods Based on Three Ordered Sample Observations from a Weibull Population, *IEEE Transactions on Reliability*, *R*-19:167–171.

Mann, N. R. (1976). Warranty Periods for Production Lots Based on Fatigue-Test Data, *Engineering Fract. Mechanics*, 8:123–130.

Maschmeyer, R. A. and Balachandran, K. R. (1980). Estimation of Repair Costs in Maintenance Agreements, *Decision Sciences*, 1:32–34.

Matthews, S. and Moore J. (1987). Monopoly Provision of Quality and Warranties: An Exploration in the Theory of Multidimensional Screening, *Econometrica*, 55:441–467.

Menke, W. W. (1969). Determination of Warranty Reserves, *Management Science*, 15:542–549.

Miller, R. A. (1973). *Making the American Motors' Buyer Protection Plan Work*, SAE Technical Paper Series No. 730586.

Moss, R. Y. (1978). Modelling Variable Hazard Rate Life Data, *Proceedings of the 28th Electronics Components Conference*, published by IEEE, pp. 16–22.

Nelson, W. B. (1988). Graphical Analysis of System Repair Data, *J Qual. Tech.*, 20:24–35.

Nguyen, D. G. and Murthy, D. N. P. (1984a). General Model for Estimating Warranty Costs for Repairable Products, *IIE Transactions*, 16:379–386.

Nguyen, D. G. and Murthy, D. N. P. (1984b). Cost Analysis of Warranty Policies, *Naval Research Logistics Quarterly*, 31:525–541.

Patankar, J. G. and Worm, G. H. (1981). Prediction Intervals for Warranty Reserves and Cash Flows, *Management Science*, 27:237–241.

Robinson, J. A. and McDonald, G. C. (1987). *Issues Related to Field Reliability and Warranty Data*, GMR-5949, General Motors Research Laboratories.

Schneider, R. W. and Tibbetts, D. R. (1983). Assessing Product Reliability from Warranty Data, SAE Technical Paper Series No. 831200.

Singpurwalla, N. D. (1987). A Strategy for Setting Optimal Warranties, Report TR-87/4, George Washington University, School of Engineering and Applied Science, Institute for Reliability and Risk Analysis, Washington, D.C.

Suzuki, Kazuyuki (1985a). Nonparametric Estimation of Lifetime Distributions from a Record of Failures and Follow-ups, *J of the American Statistical Association*, 80:68–72.

Suzuki, Kazuyuki (1985b). Estimation of Lifetime Parameters from Incomplete Field Data, *Technometrics*, 27:263–71.

Yun, K. W. and Kalivoda, F. E. (1977). Model for an Estimation of the Product Warranty Return Rate, *Proceedings of Annual Reliability Maintainability Symposium*, pp. 31–37.

8

Survey Operations Processes: The Key to Quality Improvement

DAVID R. MORGANSTEIN and **MORRIS H. HANSEN** Westat, Inc., Rockville, Maryland

1. INTRODUCTION

The goal of quality assurance in surveys is to follow operationally feasible procedures that attempt to measure what is desired and to take steps to ensure performance substantially in accordance with the survey specifications. The appropriate philosophy is to consider the total system design and attempt to achieve effective balanced use of resources. This trade-off frequently involves a choice between increased sample size to reduce sampling error and increased effort to reduce nonsampling error. While the former is conceptually easy to appreciate and relatively simple to measure, the latter is too often overlooked because of its relative intractability in being quantified. A model may be used to the extent feasible to guide this process, but in practice it must be complemented with factual information and much judgment.

We try to follow the philosophy so well advocated by W. Edwards Deming, that quality is realized by building it into the system to prevent undesirable results, rather than trying to inspect, identify, and reject poor quality work afterward (Hansen, 1987). While some inspection may be needed in any service organization, our approach is not to rely on it as a primary method for attaining our goal. Quality is designed into survey operations by beginning with the hiring of competent, well-trained staff who are imbued with high standards of accomplishment. Hand in hand with a qualified staff is the need for constant education and growth in each staff member's chosen field. Corporately, this implies strong educational and professional support programs.

The amount of quality incorporated into a survey depends to a great extent on the level of accuracy required by the client for the project and on the amount the client is willing to pay to ensure it. Some surveys call for greater quality-control efforts than others. Unlike a manufacturing operation, a service organization faces a more difficult task in quantifying the effect of poor quality and in conveying the value of more careful planning and operations to a client who may have little appreciation of such benefits (Rosander, 1985). Generally, the following discussion focuses on demanding surveys for which relatively precise and accurate results are needed.

2. SAMPLE DESIGN AND SELECTION

The quality of a survey operation begins with an overall plan that will help ensure that the sample design and size can reasonably accomplish the desired precision goals. This usually involves the careful definition of a target population and the choice of a sufficiently representative sampling frame. Area sampling, or dual frames involving list and area sampling, are often developed such that, in principle, the frame properly covers the desired target population. (By area frames, we mean the Census Bureau's mapping of the nation's geography into unique, identifiable units suitable for sampling. List sampling refers to the selection of units from an available list, as compared with random-digit dialing or area samples where no list is used to draw the sample.) Dual-frame samples involve the use of two or more sampling frames, often overlapping; used, for example, to reduce the number of sampling units that have no chance for inclusion in the sampling process or to improve the effectiveness of stratification. While a dual-frame approach may cost more in the sample selection process, it may result in reduced mean square error of survey estimates, per unit of cost, through a reduction in variance or in the number of units excluded from any chance of selection.

Frequently, choices must be made in selecting or constructing a sampling frame that is reasonably current, offers acceptable coverage of the target population, and is free or relatively free from duplication problems. Again, this may involve the use of dual or multiple frames. Often, the client's budget and schedule play a role in determining whether an existing, possibly out-of-date or incomplete frame will be used; a new, more up-to-date one will be constructed; or a dual-frame approach will be used. The greater the survey resources and sample size, the greater the relative importance of biases that can be reduced in this way.

The selection of sample size cannot be made in isolation of the needs of the consumer, the client who requires the data. This process must take into account the resources available for the survey and involves a choice between increased sample size to control the sampling error and greater effort to limit nonsampling errors. Large-scale surveys often suffer more from biases than from any lack of precision, especially in summary estimates. In smaller scale surveys, sampling error is often the more serious problem. The preferred sample size is one that strives to balance the expected precision with the anticipated accuracy in some principal statistics.

The sampling plan must include procedures for sample selection that are operationally feasible, are reasonably efficient, and have an acceptable cost. It is surprisingly easy to develop a sampling plan that cannot be implemented accurately or completely because it does not recognize the practical difficulties of field sampling activities. The plan must contain steps for ensuring that sample selection is in accordance with the specifications. Verification procedures may include comparing the sample with known totals and, perhaps, poststratification to such totals. Often, however, such totals are unavailable, and the sample estimates must stand alone. The sampling plan must be designed to permit the computation of sampling errors, without which statistical inference will be extremely limited.

3. INSTRUMENT DESIGN

A particularly important source of nonsampling error may be the survey instrument. An approach to minimizing biases introduced by the instrument relies principally on the use of highly trained, experienced instrument designers and on adequate testing. The instrument is constructed to be clear and understandable and to avoid confusion on the part of the interviewer or the respondent. A sample of respondents is asked to complete preliminary versions of the instrument so that problem areas can be identified.

There is a considerable body of research identifying special techniques for use in instrument design. Using these methods may be thought of as "building quality into the instrument." Among the principles to be applied are:

1. Begin the interview with a concise description of the survey's purpose and objectives.
2. Keep the flow of the instrument smooth; avoid excessive jumps in the pattern and the logic of the questions.
3. Use commonly accepted expressions; avoid the use of jargon or regional, specialized, or technical phrases.
4. Keep the overall instrument as short as possible; avoid the use of long questions, lists, or so-called "double-barreled" questions in which two simple questions are combined into one more complex one.
5. When respondents are asked to report on events over a period of time, a memory aid, such as a diary, should be considered to assist them. Alternatively, a bounded inquiry may be considered in which an initial survey establishes a baseline point to which a subsequent inquiry refers (for example, to avoid the reporting of purchases from an earlier period as though they occurred in a more recent period).
6. Minimize the inclusion of sensitive topics that delve into the privacy of the respondents. If such topics are fundamental to the survey purpose, they must be addressed in a forthright manner. Parental approval must be obtained where a minor is the respondent.

Pretesting and revising an instrument are fundamental to the design process. These tasks can help to verify the length of the instrument and its acceptance by respondents. Pretesting should not be limited to a handful of survey staff. For large surveys where both the schedule and budget permit, and after initial limited testing, a reasonably sized pretest involving perhaps several hundred respondents of various types is desirable. By using a pretest, instrument wording can be improved and ambiguous questions corrected or deleted.

As with all stages of the survey process, it is preferable to take the time in the design phase to prevent problems rather than to try to correct them after the fact. It is better to design the instrument to maximize respondent comprehension and acceptance and thereby avoid the need to impute for large quantities of missing data records or incomplete or inconsistent items.

4. QUALITY ASSURANCE IN FIELD OPERATIONS

4.1 Recruitment and Training

Ensuring quality in field data collected by personal interview begins with the selection and training of interviewers. To assist in the selection process, we maintain a database of several thousand people who have served as interviewers or supervisors. Using this database, choices can be made on the basis of prior performance and location. Field interviewers are generally hired by the individual who will serve as their field supervisor. Each of these steps increases the chances of finding and retaining qualified interviewers. They reduce the work of retraining and replacing staff not suited to the interviewer's role.

Clearly, increasing the interviewers' and supervisors' understanding of the survey objective and procedures will result in higher response rates and reduced respondent error. Accordingly, it has been found to be effective to commit adequate resources when developing training programs for each study. This application of the adage "an ounce of prevention" has been found to serve government survey operations as well (Diskin, 1988; Mudryk, 1988). Well-thought-out training programs familiarize the staff with the organization's philosophy, as well as all the steps they will be expected to follow during the data-collection period. Such programs provide them with ample opportunity to practice these procedures under the watchful eye of more knowledgeable staff members. Training is also used to identify new hires who may not be suitable for their intended positions.

Before training sessions, which frequently last three to five days and occasionally more, the trainee receives a home study course that the interviewer is paid to complete before the in-person training begins. The payoff from having the interviewer prepare for the in-person training is hard to quantify. It falls into Deming's category of those things that are almost impossible to measure (Deming, 1982). We have found that these preparations prevent misunderstandings and help to identify problem areas in the training process at an early stage. Accordingly, this step is judged to be valuable, especially for larger surveys where biases are more important and the costs of training and preparation represent a relatively small part of the total survey budget.

Training is often done in person. The training program may include a description of the survey objectives and a question-by-question discussion of each item in the survey instrument. Each trainee is given written materials along with the oral instruction. The program involves role-playing to provide the trainees with exposure to the kinds of experiences they are likely to encounter. Trainees are informed of typical questions that may be raised by respondents and are supplied with answers to these questions. At the completion of the in-person training, each trainee is assigned to complete the survey instruments with a few individuals (or other units) not identified with the sample. The instruments completed during this practice are examined either by the supervisor or by the home-office staff, with feedback as called for, before the trainee is permitted to begin fieldwork.

At the completion of the in-person training, a score is given to each trainee. While in-field observation of all interviewers usually is not possible, it is good practice for those rated as being in the lower one-third to receive direct observation by their supervisor during the start-up of the field interviewing. A method used occasionally to obtain feedback when direct observation is not practical is to tape record an interview. The supervisor thereby obtains a measure of the interviewer's performance on the study.

4.2 Listing, Sampling, and Interviewing

When listing for a sample of areas is involved before final sample selection, the same people who will serve as interviewers may be used to conduct the listing. At Westat, listers generally receive a one-day training program, initiated with a home-study program. The lister training involves actual in-field listing practice, usually observed or checked by the trainers. When the client supports it, in-field verification of the listing activity is conducted once field operations begin.

To ensure that the proper sampled areas are identified, interviewers are provided with detailed maps of the area to be listed. If the expected counts of housing units or other units to be listed do not reasonably match those found by the interviewer—generally, if they are off by more than a factor of two—interviewers are instructed to contact the central office before proceeding. A discrepancy can be caused by demolition or growth within an area or by error in locating the proper segment.

As the end of each decade is reached, the disparities between the previous Census counts and the actual status of the housing stock grow.

Accordingly, reliance may be placed less upon Census data and more heavily upon the use of a new construction stratum for the sampling of households. This procedure helps to eliminate surprises in field listing that may require a choice between excess field expense or subsampling with concomitant increase in the variation of sampling weights. The former results in higher cost, while the latter results in higher sampling errors, an unpleasant choice that can be avoided through the use of a new construction stratum design.

A system can be used to reduce nonsampling errors caused by incomplete listing. This procedure involves a check during the interviewing phase for structures and dwelling units missed in the listing (Waksberg, 1977) that is designed to ensure that units missed during listing are interviewed and included in the survey. The system, described in the appendix, involves a check conducted by the interviewer in a random sample of the segments and of the dwelling units. The system maintains the proper original selection probabilities, while allowing for the addition to the sample of units identified as having been missed during the listing operation.

Ordinarily, at Westat, about one-tenth of the respondents are contacted for verification of the interview process. Respondents who do not provide a phone number for a recontact are visited in person. The usual verification merely ensures that the interview was conducted. On some studies, however, important questions taken from throughout the instrument are readministered in an effort to verify the accuracy of the recorded responses.

4.3 Corrective Actions

While the survey operation is in the data-collection phase, the primary sources of errors that can be controlled stem from the interviewers and their supervisors. The challenge to the survey management staff is to separate out the system variability, which may result in some errors from all participants, from special causes resulting from specific individuals. Special causes may come from an interviewer who did not understand the training or who is not following procedures. By examining response rates for each interviewer, by scan editing interviewers' completed work to uncover item errors, and by comparing interviewers to determine the expected common cause variation, we have been able to identify and correct special-cause problems. The critical points to stress are that some errors can be expected and that survey managers must compare all interviewers to a common degree of variability if they are to uncover unusual situations. It would be poor management indeed to consider the interviewer having the lowest response rate each week "responsible" for this result.

5. QUALITY ASSURANCE IN DATA PREPARATION

Broadly, quality-assurance activities in data preparation can be divided into prevention measures and detection measures. Prevention consists primarily of more careful design of survey materials and increased effort in the training process. The aim is to eliminate steps likely to lead to errors and to provide quality training, thereby ensuring a high level of understanding of procedures by the coding/editing staff. Detection consists of inspection steps to uncover errors introduced in the interviewing or in the coding/editing process. It may involve either dependent or independent verification.

Each project will incorporate a different set of procedures, as reflected by the needs of the client. Clients expressing greater interest in ensuring accuracy will support more extensive quality-assurance activities. Needless to say, these additional steps carry an initial cost in terms of staff time to develop and to implement. We encourage our clients to invest in prevention steps, believing that these efforts are well worth their initial cost. They result in reduced errors as well as a saving of time spent later to resolve problems, such as data retrieval and rekeying, that could have been prevented (Mudryk, 1987; Cohen, 1988). More importantly, these steps reduce the chances of precipitating problems that may go undetected by the most effective of verification systems.

Some of the activities involved in data preparation are discussed in the following subsections.

5.1 Receipt Control and Field Edit

The primary objective of the receipt-control system is to ensure that instruments completed in the field arrive promptly and safely at the home office. A second objective is to identify and eliminate errors at the earliest possible point in the survey process.

This process begins in the field with scan-editing operations performed by the field supervisors. They review the interviewers' work for accuracy and completeness, and feedback with corrective action occurs immediately. Notations made on survey instruments by the supervisor are color coded for later review and summarization. Microcomputer-based files on interviewer progress and case status are maintained by the field supervisor. These are updated regularly and transmitted weekly via an automated survey-control system to the home office, where a central control file is maintained. Frequent summaries of case disposition and interviewer progress are generated for home-office staff to analyze, providing them with the data needed to correct field problems promptly. The system provides field and home-office staff with response rates and scheduling and cost information, so that prompt action can be taken where response rates are low or costs are unduly high.

When materials are sent to the home office, a multipart form is completed describing the transmittal. The supervisors retain a copy for their files, transmit one by U.S. mail, and include one with the package. The instruments generally are sent by United Parcel Service (UPS). Upon arrival at the home office, the contents are checked against the packing list and entries made in the control file to update the case status.

A second scan edit may be conducted by the home-office "data-prep" staff, but this is usually limited to data related to the case identification and its disposition code. Our philosophy is that the scan edit is best done by the field supervisor, since he or she is more likely to detect problems at a time when early corrective action can be taken. The identification (ID) information is often optically scanned from preprinted labels and moved immediately into a machine-readable format. The ID contains a check digit to help ensure accuracy in the data-entry process.

Nonresponse conversion efforts and field verification ordinarily will be conducted by home-office telephone interviewers. A sample of respondents will be contacted to ensure that instruments were completed as indicated. In addition, a sample of instruments will be reedited as a check on the field supervisor's understanding of procedures.

In many surveys, where the final stage of sampling is from lists, the original data collection is by mail with extensive follow-up of nonrespondents by telephone. The last stage of follow-up may involve personal interviews for a subsample of those nonrespondents who could not be contacted by telephone. The subsample is drawn at an approximately optimum fraction and is large enough to include the results in the sample estimates. Often, such a procedure can produce response rates of approximately 90 percent, in both household- and establishment-type surveys.

5.2 Recode

Recoding may be done to provide an indication of the degree of intercoder reliability. For example, a 5 percent sample of forms is duplicated before coding begins and is sent to independent coders. An analysis of these data can identify problems with forms that are likely to cause coding errors. If the schedule and budget permit, this recode analysis can be used in the form-design process. If not, it can be conducted for the benefit of the data users and to guide future work.

On some projects, certain data items, usually the open-ended codes involving items such as medical or occupational codes, are 100 percent verified by independent coders. Ordinarily, the regular (or precoded) fields receive a 100 percent dependent verification of the first 20 cases coded by each coder and a 10 percent sample thereafter.

5.3 Keying

We regard the best approach to keying accuracy to be well-trained personnel and careful design of instrument layout. The most accurate keyers should be assigned to the initial keying, to reduce initial errors. One hundred percent key verification may be desirable to uncover those errors that will still occur. The second key operator resolves discrepancies.

5.4 Machine Edit

A computer program is used to conduct machine edits. The editing package is used to prepare a coding manual that documents the definition of entries. An analyst defines the range and consistency checks in a series of logical statements. Most of the consistency checks are bivariate, with others involving more than two variables. The edit program is used to clean the data, and an SAS program is used to produce univariate frequency counts and cross-tabulations. The editing package allows for the definition of subject-specific edits. Discrepancies found during the cleaning process are recorded in a manually maintained log. Virtually all of the item corrections are completed manually.

6. CONTROL OF TELEPHONE AND CATI INTERVIEWING

Control of this type of interviewing follows procedures that are similar to those described for field operations. The staff selection and training process uses the same degree of care and preparation. The availability of computer support systems that can select samples and maintain data required for weighting reduces the likelihood of errors and helps to ensure that the attained results follow the planned design.

Telephone surveys offer a superior environment for monitoring and making needed corrections immediately upon detection. Supervisors are able to listen in, note any problems quickly, and take corrective action through ongoing feedback. We have recently established an office outside the Washington metropolitan area for much of this type of interviewing, where turnover of personnel is greatly reduced.

7. THE IMPORTANCE OF STANDARDIZATION

A critical concept in the improvement of quality in any operation, service, or production is the need for standardized procedures. Without the development and use of methods clearly understood and universally followed, increased and unnecessary variability will result. In the development of our sample designs and in all our procedures—i.e., recruitment, training, sampling, editing, etc.—we try wherever feasible to develop and use agreed-upon procedures.

Standardization remains important even in the presence of data-collection methodologies that are inherently more variable; for example, where measurements are made through observation or the use of a measuring instrument, as contrasted to interviewing. We emphasize in health-examination surveys the need for training, even for trained medical technicians doing operations they are accustomed to doing, to help ensure reasonable uniformity of measurements.

We have also found that substantial improvements can be achieved, where a measurement problem has been identified, by more specific training. For example, in a recent survey, we observed large variations in measurements relating to scoliosis (curvature of the spine), and detected a larger variance between primary sampling units (PSUs) than appeared reasonable. We were able to isolate this to three groups of PSUs in which measures were performed by different technicians. We will attempt to correct this problem through training procedures to be introduced into a forthcoming survey. In other similar studies in the past, we have achieved substantial reductions in measurement errors—as, for example, in measuring blood pressure, which has been recognized to be subject to high measurement variability. Despite this variability, we have, by careful instruction and training, succeeded in achieving a relatively high level of consistency in such measurements among different technicians.

8. SUMMARY

We would like to emphasize a few final points:

1. Well-thought-out preparations—be they in planning, training or design—are essential to prevent, identify, and correct unforeseen problems (King, 1987).
2. Timeliness is an important part of quality. We give timeliness a major role and make substantial efforts to make time commitments that can be met and to live up to those commitments.
3. Final critical review of statistical output for consistency and reasonableness is essential to identify problems that need corrective action.
4. We regard the estimation and provision of sampling errors, and of other measures of quality, as an important part of our procedures.
5. The commitment of management and staff at all levels to quality is perhaps the major requirement in achieving quality, all the more so for a service organization (Butterfield, 1987). We attempt to provide a working environment in which staff have a high morale and are imbued with a commitment to quality in output.

A caution is needed. Effective control of quality can, to a substantial degree, make the survey results reproducible. But, even if concept definitions are clear and explicit and effective measurement methods are available and used, the relevance of the survey results to the purposes to be served remains to be evaluated. It may be desirable to consider alternative measurement methods and attempt to understand the differences observed.

When designing a survey and deciding how to allocate resources to each step of the process, some justification is required for any effort that results in increased costs. The positive results that can be expected from quality improvement cannot always be translated directly or easily into cost savings, as they may in a production environment where wasted steps can be incorporated into an accounting process. One of the more visible and measurable quantities of survey quality is nonresponse. Although several recent papers (Cohen, 1988) have documented how the use of the steps described here have helped to improve response rates, we should not be misled into placing too great a weight on this single measure simply because of its visibility. There are other less measurable, but nonetheless positive, effects.

APPENDIX: INTERVIEWERS QUALITY CONTROL CHECK ON LISTING*

A quality check on the completeness of listings can be performed by field interviewers. This procedure was designed not only to detect and measure omissions made in the listing operation, but also to correct for them and provide an unbiased sample of missed units at the same rate as the rest of the sample.

The procedure separates the problem of looking for missed dwelling units into two parts: (a) missing units within buildings in which one or more housing units appear in the original listings, and (b) completely missed structures. For the former, a subset of structures in the sample is selected and a check made to detect all units within these buildings. If any of the units are not on the listing sheets, they are added and included in the sample. For the second group, a subsample of area segments is designated for a check of structures completely missed in the listing. All units in such structures are added to the sample.

The subsamples of buildings and segments designated for the quality check are selected at a rate such that the inclusion of all units found retains self-weighting features of the sample stratum. The procedure works as follows:

1. At each address that has been listed as a single-family home, the interviewer checks to make sure that all persons living at the address are included in the screening. If a second dwelling unit is found at the address, it is added to the sample.
2. In a subsample of multiunit addresses, the interviewer checks the listing of apartments at the addresses. If the first listed unit at the building is in the sample, the listing is checked. Otherwise, checking is not done. The check is performed by physical observation of mail boxes, doors, etc., and by inquiring at sampled households in the building.
3. The subsample of segments designated for a check of structures that were completely missed in the original listing is determined in a manner similar to the one described in Step 2. If the first unit listed in the segment is in the sample, the segment is checked for missing structures. If it is not in the sample, the check is not performed.

*Prepared by Joseph Waksberg.

Whenever a check is made, all additional structures found are added to the sample, and all units within that structure are included.

The procedure of having the first unit determine whether a check is required implies that the check is always made in structures or in segments in which all units are in the sample, in one-half of the buildings or segments in which the rate is one in two, and in one-third of the cases where the rate is one in three, etc. This retains the self-weighting feature of the sample.

REFERENCES

Butterfield, Ronald W. (1987). A Quality Strategy for Service Organizations, *Quality Progress* (December): 40–42.

Cohen, Stephen and John Morton (1988). Quality Assurance Program—Professional, Administrative, Technical, and Clerical Pay Survey, *Annual Proceedings of the American Statistical Association, Survey Research Section.*

Deming, W. Edwards (1982). *Quality, Productivity, and Competitive Position*, Massachusetts Institute of Technology, Center for Advanced Engineering Studies.

Diskin, David H. (1988). *Automating Process Controls for Surveys and Censuses*, Fourth Annual Research Conference, Bureau of the Census.

Hansen, Morris (1987). *How Large Survey Organizations Build Quality into Their Operations*, Third Annual Research Conference, Bureau of the Census.

King, Carol A. (1987). A Framework for a Service Quality Assurance System, *Quality Progress* (September): 27–32.

Morganstein, David R. (1988). *Quality Control in Survey Operations, A Discussant*, Fourth Annual Research Conference, Bureau of the Census.

Mudryk, Walter (1988). *Quality Control Processing System at Statistics Canada*, Fourth Annual Research Conference, Bureau of the Census.

Rosander, A. C. (1985). *Applications of Quality Control in the Service Industries*, Marcel Dekker, ASQC Quality Press, New York.

Waksberg, J. (1977). *National Survey of Family Growth, Cycle II, Final Report*, prepared for the National Center for Health Statistics, pp. 3–14.

9

Accuracy and Relevance and the Quality of Data

ANDREW S. LOEBL Martin Marietta Energy Systems, Inc.*, Oak Ridge, Tennessee

1. INTRODUCTION

In 1978, the Oak Ridge National Laboratory (ORNL) was asked by the Energy Information Administration (EIA) to become involved in addressing issues related to the quality of energy data. Since the time of the energy crisis in 1973 and 1974, there had been growing fears that U.S. energy statistics were not credible. The government was suspected of being gullible at best and collusive at worst regarding the data it received from industry, and industry was widely charged with manipulation and deceit. Even people experienced with federal energy-data systems expressed a generic skepticism about the integrity of data providers, the quality of data analysis, and the fidelity of data processing.

Thoughtful observers applauded the general level of interest in data quality; they read it as a long overdue and necessary prelude to establishing objectivity in energy information. But no matter how the message was read, it was apparent that the primary issue was credibility.

*Martin Marietta Energy Systems, Inc., operates the Oak Ridge National Laboratory, the Oak Ridge Y-12 Plant, the Oak Ridge Gaseous Diffusion Plant, and the Paducah Gaseous Diffusion Plant under contract DE-AC05-84OR21400 with the U.S. Department of Energy.

The Laboratory was able to bring two special perspectives to the situation. First, since the ORNL collects no primary data about the U.S. energy system, its institutional independence represents a measure of objectivity. Second, the Laboratory contributed a mix of the kinds of administrative and scientific skills that could take into account users' concerns for data quality and program managers' concerns for useful answers; that is, the ORNL could address the concerns of those who lamented the quality of energy data and those who were responsible for improving the quality of energy data.

This chapter describes some of what was learned about energy data—and about data in general—from the Energy Data Validation Program and puts forth a view of data and data quality that will alert many data users to a family of mishaps that can befall those who try to adapt public data to situations where such data are not relevant. What was learned from this program affects the way energy data are seen today, but some of what was learned about data in general has had profound implications for government management in areas well beyond energy. Two of these lessons deserve particular mention and are treated in this chapter.

The first lesson to be learned from the data-validation program is that misapplication of data—that is, taking data out of context—is a major, if not *the* major, source of error in energy-data analysis. Data are often taken literally because they are expressed in numbers. Taken within the context of the data, such uncritical acceptance means that legitimate questions about accuracy, consistency, and misrepresentation may not be raised. Outside the context of the data, this literal acceptance leads analysts into using data collected under one set of circumstances as a measure for a related, but substantively different, set of circumstances.

The second major lesson to be learned from the data-validation program is that data quality is a multidimensional problem. A recognition of the interplay between the settings in which data are collected and the settings to which they may be unambigiously applied is not new. What has been observed in this study, however, is that this interplay represents only a single, simple aspect of data quality. While accuracy problems account for many of the difficulties that were experienced in the early attempts to understand U.S. and world energy-supply figures, they do not account for all the difficulties encountered by those who were trying to understand energy use, supply, and demand. It is important to recognize in this context that numbers can be data, but they do not necessarily represent information, and that data can be "wrong" for a number of different, independent reasons.

The following sections present a brief background for the data-quality program as it evolved during the middle and late 1970s. After setting the stage for the ORNL data-validation program, the report turns to what we did (the systems approach), what we found (that the major problem with data is not misrepresentation but misapplication), and what it means (that data should be judged for relevance as much as for accuracy).

2. BACKGROUND

2.1 The Setting

In 1973 and 1974, the world economy was shaken by the Arab oil embargo. A few Americans were hurt by the energy crisis and many more were inconvenienced, but virtually everybody was surprised by it. Since "somebody" had to have "caused" the crisis, there was widespread suspicion about the motives of the oil companies and about the gullibility and impotence of the federal government. The government cited statistics and data to buttress its policies, but these data themselves were viewed with suspicion. A February 1974 report from the General Accounting Office (GAO) stated: "Perhaps the single most important problem with energy data is the question of credibility. As long as much of the reporting of data by industry is voluntary and unverified, questions will be raised as to the credibility of the data, even though the data may be entirely valid." (GAO, 1974).

By the time Jimmy Carter's presidential campaign was in full swing, the credibility of energy data had become a small scale political issue. In Congressional hearings on March 9, 1976, and later in a report issued on June 15, 1976, the GAO cited "the need to improve Federal efforts in collecting, analyzing and reporting energy data. Because the adequacy of Federal energy data continues to be a controversial subject, . . ." (GAO, March 1976; GAO, June 1976). GAO hit on the credibility issue over and over, and asserted its position that "[A] Department of Energy and Natural Resources with an independent data collection component offers the best long-term organizational solution to energy problems, including energy data problems." (GAO, June 1976).

By the time Mr. Carter took office, the pressure for energy data reform was enormous. The public challenged the objectivity of the energy information that was available (since it came from the energy industry), and it challenged the independence of the data-collection process (since that process relied on industry-provided data.) The Administration had no analytic framework that could be used to calculate what the numbers "ought to be," and it had no benchmark against which to calibrate industry-supplied data. The most sensible solution seemed to be for the government to create an energy-information validation program. However, the underlying assumptions of such a program were inherently contradictory. On the one hand, the credibility issue forced data validators to consider the possibility that data supplied by industry were suspect. On the other hand, the data validators were dependent on data supplied by industry because industry-supplied data were the only data available.

A memorandum from the Executive Office of the President, issued on August 12, 1977, just two months before the formation of the Department of Energy (DOE) on October 1, outlines the concerns of that office about the credibility aspects of energy information as it analyzed the data that DOE would need:

Memorandum from the Executive Office of the President

Subject: Credible data and information to fulfill the mission of DOE.

Major sub-issues include primary vs. secondary data, data standards, and data audit.

Developing Credible Data and Information for DOE. The basic concern in this issue is not the actual sources of data and information but rather the credibility of data and information collected from any source. Often the government has been accused of using "bad" data and information for release to the public and in formulating policy. Therefore, to fulfill the mission of DOE, credible data and information must be obtained irrespective of the sources from which they come.

Several attributes for establishing "credible" data and information can be identified. An important factor is *uniform standards*, which would include standard definitions, terms, codes, units, reporting, activities, standards for confidentiality, etc. The capability of *verifying* (*or auditing*) will enhance the credibility and utility of data or information. Primary sources of data and information are essential for some DOE purposes, such as data required for regulatory programs. From a credibility standpoint, primary sources of data are more acceptable than secondary sources. However, reliable secondary sources, which are usually more cost effective, are adequate for many purposes and they should be used where appropriate.

Regardless of whether the source of data or information is primary or secondary, it must meet the (credibility) requirements of the user.

Besides the identified organizational units that will be in the DOE, some additional functions must be established to improve the credibility of data and performance (Conklin, 1977).

In September 1977, the EIA was established by Congress as mandated by the Department of Energy Organization Act. The EIA was given the responsibility of providing objective information about current and historical energy trends, making disinterested projections of future trends, and responding to the informational and analytical needs of the DOE, the Congress, and the public.*

Proper control and documentation of the quality of its statistical and analytical information were seen to be critical to EIA's mission. Accordingly, a separate office, the Office of Energy Information Validation (OEIV), was established within EIA to independently review the methods and products of EIA. OEIV was responsible for measuring and documenting the quality and accuracy of data and analyses collected, used, and disseminated by EIA. OEIV was also responsible for reviewing the requirements for energy information and for recommending appropriate changes in EIA's collection and analysis procedures. Thus, OEIV not only acted directly to ensure the quality of EIA's product, but also acted indirectly to make EIA data system designers and operators aware of the need to improve information quality.

Data validation was not a new concept at this time. For years, the Bureau of the Census and the Internal Revenue Service (IRS) had applied rigorous and creative statistical tests to the data they collected, received, and used, in order to measure, improve, and understand the quality of data. But the limitations of these models for energy data became apparent. For the Bureau of the Census, quality of data depended primarily on sample design. For the IRS, it depended on compliance with federal regulations. These models did not address such issues as sampling bias, measurement accuracy, or the possibility that data could be misapplied. The Department of Defense (DOD) also had its own validation initiative, but it too focused on the particular circumstance of the DOD mission. For the DOD, the quality issue was not fully addressed, but it was and is

*This provision is included in the Department of Energy Organization Act, P.L. 95–91, August 2, 1977. Amended by: P.L. 95–509, October 24, 1978; P.L. 95–619, November 9, 1978; P.L. 95–620, November 9, 1978; P.L. 95–621, November 9, 1978; Section 205, Energy Information Administration. However, the mandate had existed since 1974, with the passage of the Federal Energy Administration Act of 1974, P.L. 93–275, May 7, 1974.

consistent with the DOE view. The challenge facing DOE demanded a view of data validation that was multidimensional.

At the time that ORNL became involved in 1978, almost four years had elapsed since the awareness of a need to validate energy data had surfaced during the energy crisis, and little progress had been made. Although legislation had produced the DOE, the EIA, and the OEIV, Congress was impatient for results. Proposed solutions were plagued by controversy and uncertainty, and debates about the accuracy and credibility of particular sets of information were believed to be delaying urgently needed answers to energy information problems.

In addition, the issue of credibility created an atmosphere in which the only people who could work on energy data-validation issues were those who could prove that they had had no direct experience with them. In fact, for a short time and because of a concern for possible conflict of interest, it was the policy of EIA that no contractors, or contractor staff, who had ever worked for an energy-industry company could work on an EIA data-validation project. This did not preclude working with people who were in the energy business on the government side; for example, the U.S. Geological Survey, the Bureau of Mines (now part of DOE), and the Energy Research and Development Administration (ERDA; also now part of DOE). ORNL, of course, was included in this latter exemption.

2.2 The Challenge

The challenge of assessing the credibility of energy information was massive. The unit that was being assessed was a DOE data system, which in this context represents a more or less stable, recurrent questionnaire that is periodically sent to all or a sample of all members of a particular energy activity (e.g., gasoline retailers, electric utilities, petroleum refiners, bulk terminal operators, or coal mine operators). The data system includes the questionnaire and its instructions, definitions of data elements and terms, and definitions of the universe of respondents. It also includes the data-collection processes; the sampling plan; the form; the mail-room activity; the data-processing and report-production activities; the legislative or regulatory mandate or justification for the data; the users' characterizations of how the data are used; the needs of the publications DOE issues that rely upon the data; and, of course, the database itself, as well as its operations and maintenance.

In 1978, no one knew how many data systems there were; preliminary counts were on the order of 400.* No one knew how many special-purpose energy-related databases had been built within organizations like the Federal Energy Administration (FEA), the Federal Energy Office (FEO), the Office of Technology Assessment (OTA), and ERDA; the Technical Information Center stopped counting at 2200. No one knew how long it could take to verify the accuracy of data provided by industry, but it was estimated that industry spent billions of man-hours per year generating data. †Given DOE's original staffing plans, OEIV expected to be able to provide assessments of data quality for 20 percent of the DOE data by 1983 and, at that rate (30 years for all of EIA's data), to be involved in reevaluations of the most important data sets before they ever finished the first review of the full range of data the DOE collected.

*Federal Energy Information Locator System (FEILS)—Energy Information in the Federal Government, Federal Energy Administration, Office of Energy Information and Analysis, FEA/B-76/492, Washington, D.C., November 30, 1976. In December 1973, the Federal Energy Office was created by Executive Order, and on January 23, 1974, the President directed that the Office be the focal point for energy information in the federal government. In response to the Executive Order, the Administrator of FEO created the Interagency Task Force on Energy Information to survey energy data in the federal government. FEILS represented the first step undertaken toward that goal. The FEILS data base was assembled and verified between February 1974 and July 1975 and updated in 1976. This directory is a comprehensive inventory of energy information available from 44 separate federal agencies, bureaus, and administrations conducting 279 different programs relating to energy data. As the Carter Administration prepared to come into Washington, they established an advance group called the "Transition Team." One of the Transition Team's objectives was to gather the information necessary to prepare for the creation of the DOE, a particular campaign promise of Mr. Carter's. The Transition Team counted over 260 data forms and estimated that there were over 400 energy data systems in use by the federal government (not including DOD activities and those of the intelligence agencies).

†Procedures for determining respondent burden and justifying the need for data collection are documented in *DOE Public Use Forms Clearance Procedures and Responsibilities*, Energy Information Administration, Office of Statistical Standards, Division of Forms Clearance and Burden Control, September 1981. Official government oversight is provided by the Office of Management and Budget, and DOE must provide a formal data-element-by-data-element justification and review to OMB. This is done on the form SF-83, *Request for OMB Review*, Revised March 1981.

However, despite this initially pessimistic estimate, in 1981 OEIV was able to tell the incoming administrator of EIA: "We have done most of what we were charged with doing" (Shewmaker, 1981). Although the OEIV staffing level reached only about 75 percent of its full potential, more than 80 percent of all data systems had been reviewed. The administrator was told that:

Methodologies for six aspects of data quality have been developed and generalized.
The structure of EIA's basic data series has been described and documented.
The structure of DOE's data requirements has been described and documented.
Problems have been identified and dealt with by class.
More than 500 different data forms have been reviewed.
More than 1000 field visits have been conducted with respondents and users.
Responsive contracting support procedures have been established (Shewmaker, 1981).

In reporting this success, OEIV noted that because of the size of its staff, it had found it necessary to seek help. A major portion of this outside assistance came from the Data Management and Analysis Group (DMAG) at ORNL, and it is from the perspective of having been involved in the day-to-day activities of that work that this chapter is presented.

3. WHAT WE TRIED

There are several reasons that national laboratories such as ORNL could work effectively on such tasks:

National laboratories have a reputation for successfully addressing large-scale, interdisciplinary tasks. Turning the problem over to ORNL provided a perception of progress by which the EIA could briefly allay the impatience of the energy community (Professional Audit Review Team, 1979).

National laboratories are chartered to do federal research; therefore, activity could begin without the delays normally associated with routine federal contracting. (Engaging subcontractors would also prove to be helpful.)

National laboratories have an overall obligation to sound research; therefore, objectivity and quality of thought could take precedence over deadlines and budgets.

Because these things were true, the team at ORNL was free to look at the problem of energy data validation in as broad a context as possible and to work hard to deliver quality analysis on a timely basis. We used an interdisciplinary, multiple-perspective project team (including subcontractors) with a mix of knowledge, skills and attitudes, and we set them to work in a management environment that fostered creativity and achievement. Our technical approach began with the use of systems analysis.

3.1 The Systems Analysis Approach

The systems analysis approach first gained status as an identifiable problem-solving strategy after World War II, when a spate of operations research books touted the mathematical, interdisciplinary, "find-all-the-relevant-factors" approach to problems (Ackoff, 1953; Chapanis, 1949; Churchman, 1950; Churchman et al., 1957; Hertz, 1950; Murray et al., 1947; Von Neumann and Morgenstern, 1947; Wiener, 1948). The key elements of the systems approach are as follows:

Interdisciplinary know-how as a backdrop
Total problem context as a framework
Classical statistics as a foundation

Each of these elements played a critical role in the data-validation project at ORNL.

Interdisciplinary Know-How as a Backdrop

ORNL project teams included people with expert-level skills in statistics, mathematics, operations research, engineering-economic systems analysis, and computer science, but they also included people familiar with various aspects of the energy industry. We sought diversity of practical knowledge as well as of problem-solving style. We picked people who had used data and people who had generated data. We combined pragmatists with one eye on the client's budget and one eye on the client's needs, and idealists who looked at neither budgets nor deadlines.

Total Problem Context as a Framework

This framework was established by extending the boundaries of our study to include (to the extent possible) all relevant factors. This meant including factors that simply seemed at face value to be relevant, factors that the data audience would expect to be included, and factors that lay beyond what study managers already knew to be relevant (to protect us from our own preconceptions).

These program boundaries were also established with due regard for the history of the problem of data validation and its evolution. For example, because credibility appeared to be at the center of the problem, we tried to include, within the scope of our concern, all the aspects of credibility as that concept applies to data. This meant that we would consider the integrity of the respondents, the reliability of the measurement processes, the clarity of the data requests, and the fidelity of the data-reduction procedures and processing systems and their operations and maintenance. But it also meant that we would include factors that related to the credibility of the studies themselves.

It is sometimes difficult to know how to extend these boundaries, because it is not always clear where one level of analysis ends and the next one begins. One useful method involved digging into data at one more level of detail than seemed easy and/or necessary (Underwood, 1956; Winer, 1962). In our study, this approach revealed instance after instance where data changed labels, and meaning, when tracked back through the information flow.

Another method of expanding the area of analysis involved looking at data-gathering activities that have been designed to answer a particular question, reflecting on why the question is being asked and on why each element of information is being sought, and trying to imagine how all possible answers could be interpreted. If one of the possible patterns of response is difficult to explain or produces ambiguous (or counterintuitive) inferences, then there might well be some additional data elements that ought to be collected.

By looking beyond the written or stated purposes that the data systems were said to be addressing, and what we would otherwise have naively assumed they were addressing, we found instance after instance where the DOE had been given senseless data simply because users had not known what to ask for. Many of the people in industry with whom we talked resented the burden of government data collection, but were most outspoken about what they called manifest lack of understanding by the government of the pragmatics of how the energy sector of the U.S. economy works. This same issue has been found to apply to a number of federal government nonenergy domains, such as reliability-centered maintenance data, data for all sorts of simulations and gaming areas, risk and capability assessments, and environmental data. Respondents try to answer the data requests they are given, but when the data requested bear little or no relationship to how they operate their businesses, they are less than patient with criticism of their responses and more than patient with questions about which data ought to be collected and how.

As stated above, however, our concerns were not only with what other people wanted or expected. We wanted to ensure that our own judgments about credibility were themselves credible. Therefore, we included in our study several external checks on ourselves as well as on the data we were examining.

On a selective basis, we made duplicate analytic assignments, using known biased data sources as a control, so that we could account for the differences in results. In addition, whenever we approached a new issue in a data system, we sought advisors or experts to whom we could assign a kind of adversarial role to help us identify both the dimensions of that issue and the dimensions of the probable arguments about that issue. As an example of how we tried to provide a check on ourselves in this way, in studying industrial energy-conservation data systems, we included an overview of the federal government's own energy-conservation data system (Koopman, 1978). We did this so that we would not "accidentally" find ourselves later telling the government that it should cite a particular data-system "flaw" or recommend some particular data-system improvement, and then itself be vulnerable to the charge, "How can you ask industry to do so and so when the federal government's own data system does something else?" (Brais et al., 1974).

3.2 Classical Statistics as a Foundation

Using the systems analysis approach, we began by formulating a statistical model that would enumerate the sources of error in a DOE data system:

1. Specification Error—difference between desired attribute and measured attribute
2. Selection Error—including coverage or frame-construction error, sampling-design error (if applicable), and nonresponse error (including treatment of imputation)
3. Respondent Error—misunderstanding of questions, measurement error, estimation error, inadvertent errors, and falsifications
4. Processing Error—coding error (if applicable), keying error, and systematic processing error (related to such things as programming, editing, and aggregating)

SYSTEM FLOW

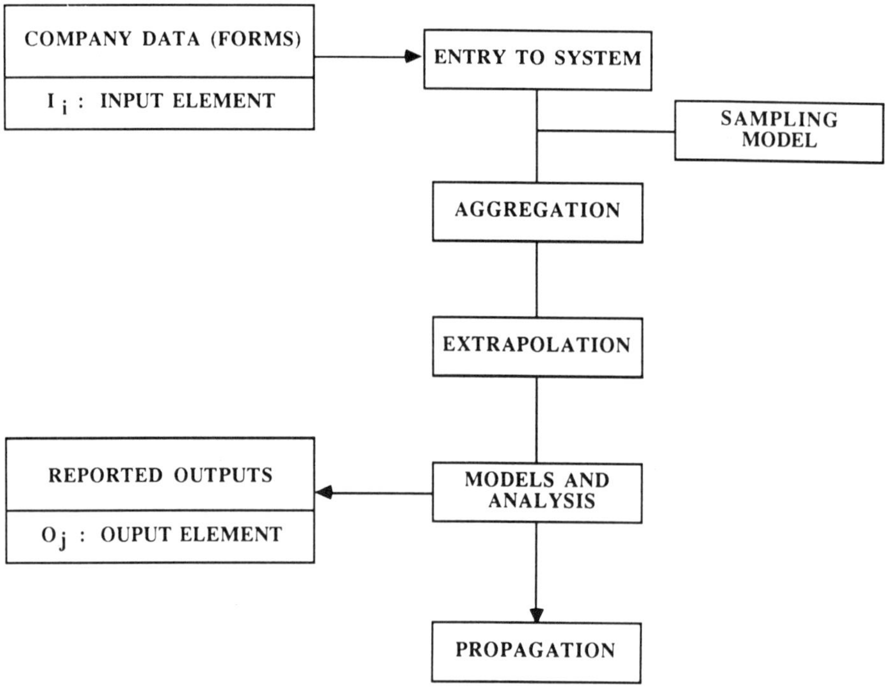

Variance on Input $s^2 I_i = s^2 I_{iA} + s^2 I_{iB} + s^2 I_{iC}$

A relates to accuracy of measurement.
B relates to clarity of instructions.
C relates to quality control by companies.

Variance on Output $s^2 O_j = s^2 I_i + s^2 P_{i/j}$

P relates to propagation errors in data system.

Figure 1. Generic error model.

We then characterized a generic error model (Figure 1) that treated total error as an appropriately weighted sum of the variances of the components.

With this generic error model, data quality analysis could run in either of two ways:

First: Determine s^2I_i, and then propagate s^2I_i to determine s^2O_j (i.e., measure input variance and observe distortions during data-system operation, then state the impact on and accuracy of O_j).

Second: Determine user needs (s^2O_j), and then propagate backward to find the required s^2I_i (i.e., determine user needs, offset distortions during data-system operation and set accuracy standards for I_i).

As our study progressed, however, we found that although a classical statistics approach was a useful place to start, it was at the same time both too narrow and too broad.

It was too narrow in the sense that although classical statistics can enumerate all the sources of error, its applications almost invariably enumerate only those errors that are associated with data collection and data processing; they seldom, in fact, enumerate sources of error associated with the "informational content" of the data. Classical statistics also fails to rigorously address variance due to specification errors, which were found to be a predominant source of errors in many energy data systems. The classical statistical approach recognizes misspecification as an error component, but it is at a loss to quantitatively assess it. For example:

What is the contribution to error in an industrial energy-conservation data system that measures not conservation, but changes in industrial energy consumption? The Voluntary Business Energy Conservation System (VBEC) and the Industrial Energy Efficiency Improvement Program (IEEIP) both call their data systems "conservative," but both collect data on consumption, and neither adjust the consumption figures for changes in production levels or for changes in heating or air-conditioning requirements from year to year. While these data are surely related to conservation data, they are not conservation data per se (Brais et al., 1974).

Or, what is the contribution to error and variance in a gasoline-consumption data system that measures gasoline tax revenues? The Federal Highway Administration used Form PR-551M to collect data on gasoline sales tax revenues by state, and then it imputed gasoline consumption by state. Such a system is roughly right, if average per-vehicle inventory changes from month to month are minor, or net to zero (which they may not be in times of gasoline shortage), and if purchases in each state are assumed to be consumed in that state, or to net to zero with adjacent states (which they do not generally do, for a variety of reasons). Again, the data seem related to their stated referent, but they are of unknown precision (Lancaster, 1971).

The classical statistics approach was too broad in that it had no capacity to identify individual data points as suspect. This becomes a key weakness if data credibility is to include the conventional concept of misrepresentation. (For example, how does one distinguish between the randomly occurring observation that is "two sigmas above the mean" and the respondent who falsely reports that his plant used that much energy? Which outlier is genuine? Which "observations" are outliers?)

These inadequacies were addressed by compounding the rigor of statistical inference in two directions.* On the one hand, we dealt with the problem of narrowness by broadening the logical foundation to consider the full range of specification error: What did data users ask for, what were they given, and what should they have asked for? We developed a comprehensive, general approach to assessing the quality of a data system that explicitly included the concepts of accuracy, compliance, and relevance. We called this formulation the Data Validation Methodology (Figure 2), and we demonstrated and refined its effectiveness in several case studies (Evaluation Research Corporation, 1979; Oak Ridge Associated Universities, 1979; Transportation and Economic Research Associates, 1980; Transportation and Economic Research Associates, 1981; BDM, 1980).

On the other hand, we dealt with the generality of classical statistics by narrowing the approach, to use the power of statistical analysis to

*These two methodologies were brought into full application and refined by the ORNL Data Validation Program, but the seeds for their development were sown several years earlier. The earliest reference found for data validation per se occurs in Leuba, H. R., Validity of Field Maintenance Data, *MAC Reliability and Maintainability Panel*, Arlington, Va., March 23, 1966. (Report of work done under U.S. Air Force contract at ARINC Research Corporation, Annapolis, Md.) And the earliest reference found for automatic data editing in the context of data validation is: Johnson, T. G., J. I. Naus, and R. T. Chu, *Data Validation Handbook*, GEOMET, Inc., Rockville, Md., GEOMET report No. IF-68-1, November 15, 1971, under contract to the U.S. Navy.

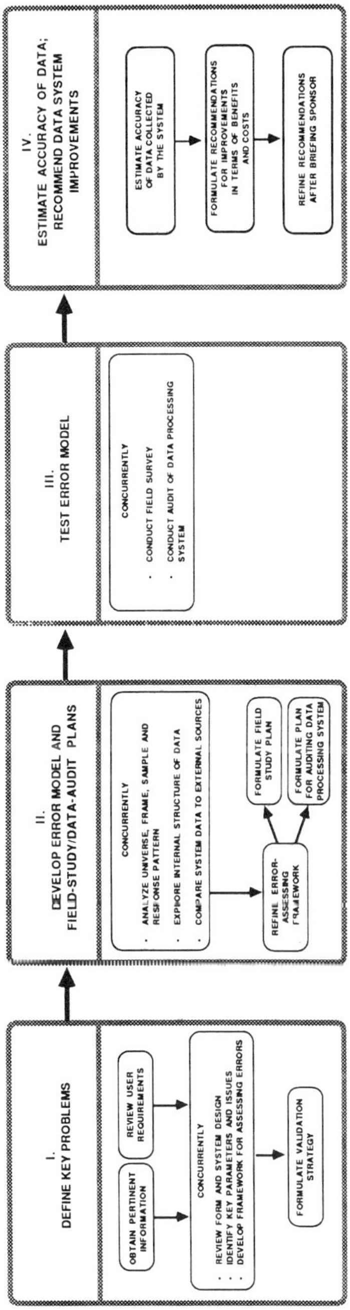
I.
DEFINE KEY PROBLEMS
OBTAIN PERTINENT INFORMATION
REVIEW USER REQUIREMENTS
CONCURRENTLY
. REVIEW FORM AND SYSTEM DESIGN
. IDENTIFY KEY PARAMETERS AND ISSUES
. DEVELOP FRAMEWORK FOR ASSESSING ERRORS
FORMULATE VALIDATION STRATEGY
II.
DEVELOP ERROR MODEL AND FIELD-STUDY/DATA-AUDIT PLANS
CONCURRENTLY
. ANALYZE UNIVERSE, FRAME, SAMPLE AND RESPONSE PATTERN
. EXPLORE INTERNAL STRUCTURE OF DATA
. COMPARE SYSTEM DATA TO EXTERNAL SOURCES
REFINE ERROR-ASSESSING FRAMEWORK
FORMULATE FIELD STUDY PLAN
FORMULATE PLAN FOR AUDITING DATA PROCESSING SYSTEM
III.
TEST ERROR MODEL
CONCURRENTLY
. CONDUCT FIELD SURVEY
. CONDUCT AUDIT OF DATA PROCESSING SYSTEM
IV.
ESTIMATE ACCURACY OF DATA; RECOMMEND DATA SYSTEM IMPROVEMENTS
ESTIMATE ACCURACY OF DATA COLLECTED BY THE SYSTEM
FORMULATE RECOMMENDATIONS FOR IMPROVEMENTS IN TERMS OF BENEFITS AND COSTS
REFINE RECOMMENDATIONS AFTER BRIEFING SPONSOR

assess the accuracy of data sets. Since work of this kind begins with a detailed description of the data set (including its parameters and moments, a histogram, and error tests), and since, in today's world, that kind of work is generally performed by computers, we called this work "automated data editing," even though it was a concept that entailed much more than mere editing.

3.3 Data Validation Model

The primary task of the Data Validation Methodology was to assess the quality of data systems. The phrase *data system* refers to the process of collecting data, as distinct from the numbers per se. It deals with the method of data collection more than with the results and addresses the performance of the whole data-collection system (forms and instructions, sample frame, data processing, respondent bias, and user inferences). The automated data-editing approach, on the other hand, assessed the quality of data sets. The phrase *data set* refers to data that have been collected, processed, and made available to users. In this approach, the focus is on the numbers, rather than on the underlying phenomena, the measuring process, or the use of the data (Liepins, 1980; MAXIMA Corporation, 1980). That is, it looked at the structure of data that had already been collected and explored outliers and other inherent evidence of distortion. In a limited but nonetheless accurate sense, our Data Validation Methodology was concerned with the potential accuracy and meaning of the data collected by EIA, and the automated data-editing work was concerned with the realized accuracy of the data collected by EIA.

The guiding concepts of the Data Validation Model were meaningfulness and determination of data accuracy.

Meaningfulness. This concept involved determining, from interviewing leading users, which data were used for which purpose. Meaningfulness led to setting priorities, which, in turn, led to deciding which data elements would be selected for validation.

Determination of Data Accuracy. The pivotal tasks involved, first, the generation of an error "model" (actually, the formulation of sources of error in the data elements that had been selected on the basis of meaningfulness); and, second, the checking of company records to arrive at an estimate of the accuracy of the data as submitted by the company.

The analytic approach underlying the Data Validation Methodology derived from one fundamental fact: requirements for information about the phenomena of interest and the phenomena themselves are in a state of flux. However, the significance of these dynamics was not directly recognized when we began our first studies. One consequence of this state of flux is that, while a data survey may indeed have been properly designed at one time (some as long ago as the 1930s), user needs may have changed significantly since the initial design. Another consequence is that the phenomenon itself (the industry structure and/or the operations) may have changed since the data collection was last (if ever) tested. We realized, therefore, that the Data Validation Methodology must begin with a review of the requirements for the information, a study of the available information about the data-collection system, an examination of the dynamics of the phenomena the data refer to, and a review of such alternative information sources about the current and past status of the phenomena of interest as may be available. Similarly, our recent work in developing computer-system modernization methods has resulted in a methodology that closely parallels the energy data and model-validation method described here.

Based on these conclusions, we were able to identify three major components that, though they may vary from study to study, should be considered in every study: (1) the user's requirements, (2) the data-collection/data-acquisition process, and (3) the nature of the events being reported on (and the interaction between those events and the reporting processes).

These approaches to data (model or computer system) validation were developed during the first year of the program by the Data Management and Analysis Group. In addition to these, however, because our management philosophy did encourage a creative environment, we were able to pursue several avenues of research at once, some of which were included in and some of which worked in concert with the more classical approach already discussed.

Our initial efforts included case studies that focused on validating the accuracy and suitability of selected data sets and data-collection systems (ORNL, 1980). Because the goal of this work was to uncover disparities between data elements as specified and data events as reported, we relied heavily on field work for this kind of validation study. This approach included conversations with people in industry who fill out the energy data forms that DOE sends them, conversations with the people in DOE who send out the forms (the data-system operators), and conversations (as well as other kinds of observations such as a review of written materials) with data users.

Although the data users provide the ultimate rationale for validation, and the data-system operators control the means for data-system improvement, the respondents (the data providers) are the ones who have the best, and sometimes the only, hands-on experience with the phenomena and the data. Data users lament alleged data inaccuracy and state the reasons for needing data, and data collectors are responsible for collecting the data, but neither group necessarily knows what the data mean or how the data relate to the events described. (Recent Paperwork Reduction Act regulations require a "purpose for collection" explanation to be available on all government forms and questionnaires.) We frequently encountered data-system operators who had been operating federal data-collection systems for 15 to 20 years, but who had never been inside even one of the facilities about which they collected data.

Only the data providers have a perspective on the events, but they seldom know how the data will be used. We discovered a wide but invisible gap between data "needs" and data. The gap was wide because, for most data systems, there was no oversight of the connection between the phenomena and the data application. That is, since there was no one with a view of both sides of the process, whatever correlation there was existed by virtue of the structure of the data system and a common vocabulary. The gap was invisible for the same reason. Again, there was no one with a view of both sides of the process, so whatever distortions existed were hidden by an apparently valid data-system structure and a seemingly common vocabulary. Field work made these gaps visible and thereby helped to close them.

Most of the field work was performed with the assistance of subcontractors who were not in the energy business, but who had contacts and acquaintances in the business. The fact that the field representatives were often known to the respondents speeded the process along, since it established an atmosphere of friendliness and trust and made it easier to get the necessary information.

We used both free-form and structured interviews in our field work. We relied on the structured approach to make sure that we covered all the issues we knew about (e.g., bias, burden, timeliness of data, and definitional uncertainties), and we relied on the free-form approach to let the conversation turn to topics that were of concern to the respondents, but about which we had no prior knowledge (e.g., measurement conventions when a shipment is being loaded, internal rules of thumb and correction factors, and inconsistencies hidden or caused by idiosyncratic definitions).

The field work was augmented, both in advance—to ensure adequate preparation so that we could recognize important occurrences when we encountered them—and after the fact—to integrate, interpret, and draw inferences from what we found. These efforts included a broad range of activity:

We undertook legislative research to uncover the intent of Congress in authorizing the collection of particular data elements. This research was always interesting, but seldom produced explicit rationales. We soon learned that most energy data are collected on the basis of general authority (e.g., "monitor as required").

We reviewed literature about the phenomena under investigation to gain familiarity with the phenomena and to appreciate possible data uses.

We used statistical methods to assess the impact on data variance of imprecisions in sampling plans. For example, applying the Data Validation Methodology, we would observe whether or not sampling was relevant. Since sampling implies collecting data from less than the whole population, sampling would not be relevant if the purpose of the data was regulatory compliance with respect to a particular respondent, and many energy data (for Federal Energy Regulatory Commission and Economic Regulatory Administration, for example) were for regulatory compliance purposes. On the other hand, if the purpose of the data was to describe a process, then sampling would be relevant, even if the data collection was first designed as a census. If sampling was relevant, then the automated data-editing project team used statistical methods to evaluate the interaction between sampling-related bias and the distributional structure of the data that were collected. (For example, could bimodality, skewness, or truncation be attributed to sample design, or was the distortion a true reflection of some characteristic of the underlying distribution?)

We developed deterministic edits for data elements that were data-set specific. The deterministic edits ranged from simple to complex tests. The simple tests included such edits as checks to see if the parts added up to the whole or to see if units of measure were plausible. The complex tests were phenomenon-based and situation-specific. Examples of this included checks that alleged fuel flows were within the delivery capacity of the systems supposedly involved or that the deducible production efficiency was compatible with historical trends or design specifications. In doing this data editing, we were blending knowledge of the phenomena involved with knowledge about how to do analysis of the data that were available. That is, we were using automated data-editing techniques as a way to close the gap between the phenomena out in the field and the users' needs for data that mean what they seem to mean and are measured in a reasonably accurate way.

(This data-editing project serves as an excellent illustration of the value of an interdisciplinary team, where levels of practical experience blend with types of technical knowledge. For example: It is as unlikely that a journeyman statistician or computer programmer would know enough about steam-boiler operation to be able to anticipate negative electricity generation (ORNL, 1980) as it is that a steam-boiler engineer would be familiar enough with statistical analysis techniques to be able to use Dixon's ratio in a computer program to single out which data reporting positive electricity generation were outliers.)

In the automated data-editing work, we looked at sets of data as raw material for testing and developing a methodology to be used in identifying outliers. We used both naturally occurring data sets and data sets with known (introduced) errors to test new statistical tools, or new applications of available statistical tools, for identifying outliers and imputing missing values. As a result, we were able to understand how the state-of-the-art techniques in statistical inference were applicable to the assessment of energy data quality.

One other approach that worked well and needs to be mentioned briefly was the use of exploratory analysis, a concept that refers to that variety of data analysis described in John Tukey's book *Exploratory Data Analysis* (1987). It is fundamentally statistical analysis that draws on the perceived structure of the events being studied, rather than trying to impose on those events an arbitrary statistical theory. (In this regard, exploratory analysis has the robustness of nonparametric statistics with a perspective of heuristic research as opposed to hypothesis-testing research.)

In addition to these approaches, several elements and procedures of an organizational nature were established (both initially and as the work progressed) that contributed to the success of the study:

We were able to augment our resources with skills and staff labor from outside sources by employing the competitive procurement process; we could pick the most qualified people and rely on the competition to ensure that they would work hard.

We used a prespecified structure for reports and briefings. Working out these outlines early in each task helped to identify issues and focus the activity. This approach also enabled us to review all reports quickly and gave the team an ongoing feeling of accomplishment.

We established subject matter as the primary and most efficient means of understanding individual data forms, realizing that energy data could be systematically validated (i.e., reviewed and documented) on a subject matter, rather than a form-by-form, basis with considerable economy of scale. The topical studies were much more complex (and expensive) than were the single-form studies, but they were also more comprehensive (and more satisfying) and could cover a number of individual forms.

We allocated study emphases to areas where data improvements were feasible, as opposed to just interesting. The energy-data-validation program began with a taxonomic approach that assumed that each data form would be evaluated and that the sequence of validation studies would follow some orderly procedure. This was not as arbitrary as addressing the forms in simple numerical sequence, but it did begin by addressing them in market-sector sequence. As soon as the first preliminary results were available and tentative assessments were made of how the results might be used, it became clear that there should be a priority sequence based on the potential impact of probable data-system changes. This sequence was immediately developed, maintained, and followed.

After the first case studies were published, the emphasis moved away from field work and related activity; in part because field work was expensive, but also because what had been learned from the case studies guided us to our next steps. The validation effort moved, therefore, from assessing data-system performance with a view toward improving data quality to assessing data-system performance with a view toward characterizing how sound and authoritative the data were for the intended uses. It shifted from having the audience for the data-validation report be the DOE group responsible for designing, operating, and maintaining the data-collection system and those responsible for disseminating the data, to having the audience for the data-validation report be the data user. It became de rigueur to write warnings specifying the degree of error inherent in the data and the limitations on their applicability.

One of the most persistent difficulties throughout the project was that of identifying data users. Even when the focus was on improved quality, the improvement was supposed to be vis-a-vis intended data use. But intentions were seldom explicit. Frequently, the only way to understand intentions was to infer them from the language of the legislation that authorized the collection of the data. But this was an imprecise process, since the language was invariably either vague or general, or both. The other alternative was to "backcast" the intended purpose of the data from the uses to which they were actually put, but there we had the problem of finding uses and users, of weighing the relative importance of different uses, and of identifying users who had considered the data and determined they could not use them. We eventually finessed the issue in this setting and solved it in the broader setting.

We finessed the definition of user in the data-quality context by characterizing data quality with respect to the uses for which the data were reasonably valid. That is, we pretended that there was no specification error and said, "If these are the data that are wanted, then these are the settings to which they apply, and this is how accurate they are."

We solved the problem of defining a user in the broader setting by undertaking a review of the data that would describe a closed set of phenomena. This was also called a *review of requirements* (for information) and, under its cover, we circumscribed the parameters relating to a particular energy topic (e.g., coal-production data, petroleum production, feedstocks and crude-oil imports, electricity distribution, and cogeneration). We developed a model of the phenomena identified possible data points, and evolved a minimum set of information requirements that would categorically describe the operation of the phenomena and their interdependencies. For this approach, it did not matter that we could not define a user, because the data set was complete. Since the data set covered the whole phenomenon, it had to meet the needs of any user who was concerned about some aspect of this closed set of phenomena.

The final step in our formal approach was integration, blending what we learned and applying the information to all our studies in ways that would improve the process of data validation in general.

Reviews of requirements, system analyses, and integrations all addressed the characteristics of data. A review of requirements addressed what the characteristics of the data ought to be, a system analysis addressed what the characteristics of the data in a data system actually were, and an integration addressed the options for improving the characteristics of data. The review of requirements, generally conducted first, shaped the system analysis by identifying the kinds of data characteristics that were important regarding a given use to which the data might be put.

The integration, by definition, was conducted after the requirements review and the system analysis. Using these techniques as a foundation, the major data series and topics were analyzed.

4. WHAT WE LEARNED

In the four years from 1978 through 1981, the Data Validation Program examined the accuracy and meaning of the data produced by 19 major sets of data-collection forms and systems. EIA estimated that this work dealt definitively with the validity (and credibility) issues of at least 80 percent of the energy data collected, used, and/or distributed by the federal government.*

Simply put, our charge had been to look at the energy information EIA collects, uses, or publishes, and ask:

Who needs the information? For what?
What data do they really need?
How good does it have to be?
How good is it?
Does it need to be improved?
How can the improvement be accomplished?
Are the analytic tools or models appropriate?
Are data available to feed them?
How can the amount of data collected be reduced to a minimum?
What is that minimum?

Data users have asked questions like these for years, but up to this point no one had codified a reasonably complete (i.e., multidimensional) methodology for answering the questions. At first glance, the problem of improving data quality seems to be a statistical question—an issue of sample size, sampling error, and response bias—but when we got to the field and talked to the people who fill out energy data forms, we found that

*When the acting director of the Quality Assurance Office of Statistical Standards, EIA, reported the status of this work to his peers, he listed 82 "basic series" data systems and declared validation efforts for virtually all of them as "close to being complete." (Forty-seven of the basic series either did not require validation or their validation studies were complete or under-way. The bulk of the other systems were believed to be fairly accurate in any case.) Hale, D. R., American Statistical Association, Chicago, Ill., October 22, 1981.

the issue was much more than statistical. To be sure, valid data rest on obtaining information from an appropriate sample of respondents, but valid data also rest on the clarity and sense of the questions on the data-collection form. One of the most important conclusions of our work to date is that a precise statistical calculation of probable error is almost (if not altogether) irrelevant compared to questions of bias. It is not that we found rampant misrepresentation, nor that one does not need to deal with the probability that nonrespondents come from a different segment of the population than do respondents. Rather, we found that the energy sector of the U.S. economy is so heterogeneous that data provided in answer to a first-draft question may be full of idiosyncratic responses. That is, there is much more room for error in interpretation or applicability or variation in definition than there is for uncertainty due to random variation. Thus, where conventional statistical wisdom has focused on survey design, error profiles, and homogeneous classes and covered interviewer bias, coding error, and nonresponse coverage when it studied system errors, we did that; but we also had to go beyond classical statistics (because the sources of error are outside the sample design) and beyond classical auditing (because that precision is not worth the cost) to develop the methodologies and procedures discussed in the previous section to deal with open-ended variation in interpretation and heterogeneity in populations.

This is what we found:

> *While every data-collection system offers a host of opportunities for random error, the systematic errors in large-scale data-collection systems are much more significant than are the random errors.*

In May 1981, when OEIV compiled a list* of all the recommendations that had been made in the formal reports published by the energy-data-validation program, there were 99 recommendations in all. Virtually all of these recommendations dealt with systematic errors between what the data were described as collecting and what they did collect. As can be seen in Table 1, which summarizes a categorization of the 99 recommendations, 89 percent of the recommendations dealt with meaning or bias-related errors and only 1 percent dealt with measurement-related errors. Only one of the 99 recommendations dealt with random errors in a data system.

*Roemer, D., *Printout, OEIV Recommendations*, DOE/OEIV internal documentation, (GC5.STUDY.TITLE), May 8, 1981, unpublished. This pair of computer listings included the gist and status of all recommendations included in OEIV reports from October 1, 1978, through May 8, 1981.

Table 1. Categorization of OEIV Recommendations

Category of Problem	Number of Recommendations	Percent
Compliance	1	1
Measurement	1	1
Data processing		9
Input errors	0*	
Procedural efficiency	9	
Data clarity		42.5
Presentation to users	24	
Description to respondents	18	
Data coverage		46.5
Adjust frame	16	
Add data elements	20	
Drop data elements	5	
Relate to other systems	5	
TOTAL	99	100

*Some data-processing errors were found, but the corrections were so straightforward (and the impact was so minor compared to other sources of error) that the need for corrective action was not given the status of a formal OEIV recommendation.

Source: D. Roemer, *Printout, OEIV Recommendations*, DOE/OEIV internal documentation (GC5.STUDY.TITLE), May 8, 1981 (unpublished).

> *Conscious bias or misrepresentation are possible, but are likely to cause massive distortion under only one set of circumstances (when there is a low-risk incentive to distort).*

Not even one OEIV recommendation dealt with the possibility or existence of conscious bias or misrepresentation. We know that such misrepresentation can occur—considering just one example, the interaction between inventory tax or customs duty or tier pricing and reported stocks or sources of oil, demonstrates that incentives for such activity exist—and cases of federal prosecution for such offenses are reported from time to time. Indeed, we began our work assuming this to be true. However, our study produced no evidence of it and did reveal a host of practical institutional barriers to such conscious misrepresentation.

In the first place, most of the data DOE collected and used has no direct consequence for the respondent other than the burden inherent in providing it and the penalties associated with negligent or fraudulent reporting. In the second place, because the reported data have to be verifiable, if challenged, by direct reference to company records, a company bent on providing false data would have to keep two sets of books, one with which to manage its own affairs and one to use to provide data to DOE. It is possible to do that, of course, but the cost of two sets of books is likely to be greater than the economic "benefit" of providing directly false data. Third, the risk of discovery is prohibitively high. To provide intentionally false data, many people within the company would have to cooperate in an act that they knew was illegal and/or unethical. Since all of them would be vulnerable to the pangs of conscience of any one of them, it would be difficult to achieve the level of conspiracy necessary to carry out a data deception.

In the course of our study, we interviewed several thousand data respondents, and after examining the economic, physical, and regulatory structures of both the petroleum and the coal industries, we believe that competition and the profit motive have interacted with regulation to modify what the energy industry does, thereby modifying only indirectly what is reported to the DOE. In other words, the effect has been to modify what industry *does*, and by so doing to modify what it reports. We found virtually universal accommodation between industry practice and the subtle and direct incentives in regulation.

Sometimes, this interaction with and response to regulation produces curious, otherwise uneconomic, activity. For example, if the law says that U.S. crude oil for U.S. users has to be carried in U.S. ships, then it is carried in U.S. ships, even though it costs more to use U.S. ships than to use foreign ships. However, sometimes it is cheaper for California refiners to send the U.S. crude oil from Alaska to Japan in foreign ships and to buy foreign oil for U.S. use than it is to buy Alaskan crude oil for U.S. use and ship it in U.S. ships.

> *The single most significant source of error in data analysis is misapplication of data that would be reasonably accurate in the right context.*

As stated above, we conclude that there is a complex interaction between the economics of energy operations and the incentives that flow (both intentionally and unintentionally) from energy regulation. This interaction is manifest, first, in how business is accomplished, and then in what is reported. We also conclude, however, that the interaction among regulation, operations, and economics is so complex that the reported data should never be taken at face value. The best that can be expected is a close approximation to the phenomena of interest.

A detailed examination of the data-clarity recommendations or of the data-coverage recommendations (in Table 1) illustrates the fact that data cannot just be taken off the shelf and used as-is. An analyst must understand what was measured before he or she begins to use the data.

No data system that was examined could be safely taken at face value. Every data system studied yielded at least three data-clarity or data-content recommendations, many of which referred to problems that could easily invalidate the data. There are, for example, conservation systems that measure consumption; consumption systems that measure sales; inventory systems that measure distribution; refined-products systems that include crude oil; crude-oil systems that exclude imports; and confusion between custody and ownership, between receipts and shipments, and between "on hand" and "available." There are, in short, many data systems that do not measure what their names imply that they measure. Analysts cannot use data from these systems unless they thoroughly understand the underlying phenomena—and even that does not always mean that the data can be made useful.

5. WHAT IT MEANS

Energy data, and the quality of energy data, became interesting topics when the real price of energy began to increase after 1973 and the real price of almost everything else rose as a consequence. Energy data became an interesting topic, a topic where belief in the accuracy and credibility of the government's energy numbers was important, in large part because those numbers had an impact on everyday lives. People would be cold, run out of gas, or pay more for energy products than they could afford as a consequence of those numbers.

As energy data became more interesting, questions were raised about the motives, or at least the effectiveness, of the institutions most directly involved. A skeptic could challenge both the conclusions and the motives of an analyst simply by pointing to the fact that data came from the most suspect source: the energy (read "oil") companies themselves. It did not matter to this argument that the energy companies had no conflict of interest with regard to the particular data-dependent decision at hand; it was sufficient simply to point out that they had vested interest in energy per se. (For example, during the height of both gasoline shortages, press stories characterized DOE as being suspected of bias and collusion and the oil companies as having caused the shortage when, in fact, the decision being challenged was the allocation fraction or state-set-aside figures—neither of which was under the control of the oil companies. The oil companies may well have benefited (or not) from the gasoline shortage, but only the most shallow thinking would assume that the oil companies manipulated their inventory and demand data with the intent to differentially influence state allocation fractions and set-aside amounts.)

5.1 Credibility

As stated at the beginning of this chapter, the primary issue in the mind of the public concerning energy data is credibility, and the primary issues of credibility regarding all kinds of data have to do with the perception of the user. These can be expressed in a weighting of the following perceiver-specific possibilities regarding energy-data questions:

1. The answer is or is not what was expected.
2. The answer is or is not what one wanted (or does not matter).
3. The answer does or does not serve the interests of the data provider (or does not matter).
4. The data system does or does not faithfully represent the underlying phenomenon.

This formulation confirms the fact that credibility is a derivative characteristic of data, not necessarily an intrinsic one. Lack of credibility may derive from the data because the data may seem incredible even though they are true, or credibility may be conferred upon data that may or may not deserve it. The point is that there is more to credibility than the honesty of the data provider and the fidelity of the data-collection system. Credibility also depends on the context in which data appear and are used.

However, neither of these two aspects, the traditional accuracy aspect or the contextual aspect, represents a single dimension of information. The accuracy aspect includes, for example, specification errors and measurement (precision) errors, as well as processing errors and biased reporting. The contextual aspect includes the relevance of the data to the context and the biases of the person conferring or withholding credibility.

In the data systems we studied in the energy-data-validation project, the contextual aspects of credibility were much more important than were the accuracy aspects, and they dwarfed the honesty component of the accuracy aspects. In our experience, occasions of misapplication were hundreds of times more likely to be the cause of misleading energy analysis than were occasions of misrepresentation.

One of the things our study revealed is that what might be called errors of analysis were much more important than errors of bias. (See Figure 3—dimensions 1 and 5 vs. dimensions 4 and 6.) According to this model, the public was "right" to withhold credibility from energy data analyses in the middle to late 1970s, but their justification for doing so derived neither from their own biases (kill the messenger if you don't like the message), nor from misrepresentation by the data providers, though both feelings were perhaps understandable. The fundamental justification for withholding credibility from energy data analyses in the middle to late 1970s was that the analyses themselves misapplied reasonably accurate but fundamentally inappropriate data. The phenomena involved were simply not well enough understood to support the massive transference of data collected in a quasi-descriptive milieu to analysis in a quasi-regulatory milieu. Not only did the data change meaning in this transfer, but the underlying events changed in response to the changed milieu.

The distinction between accuracy and credibility (or between measurement precision and data relevance) may be difficult to understand. A reason for this may be that the word credibility is a concept that covers both disbelief because the implications of the analysis that uses the data are unacceptable and disbelief because the data themselves are suspect.

A few examples may help to clarify this distinction. It is obvious that reported sales may be a biased or inaccurate number if the respondent has to pay tax on reported sales, and it is obvious that a reported sales figure is not the same thing as consumption. (It might even be obvious that even if sales were measured accurately, they are not the same thing as consumption.) But it is difficult to illustrate the general principle that measured sales may not be the same thing as measured sales—that is, to illustrate that even when the same word is used, the underlying referents may well be very different. Data may measure sales to include interdivisional transfers and then be reported (or used) as though there were no such thing as interdivisional transfers, or sales may represent net returns (or not) or be based on different time periods (does one include billed sales or only paid sales?) or geographical regions (are extraterritorial sales credited to the region of the seller or the buyer?).

The following examples further illustrate the point:

Conservation isn't conservation. The VBEC reported on industry's efforts to conserve energy—but it did so by reporting consumption over

Aspect I: Accuracy

Dimension 1 Specification error

Examples:
- measuring the wrong thing
- sampling plan errors
- ambiguity in defining data elements

Dimension 2 Measurement precision

Examples:
- noise in recording
- errors in meters

Dimension 3 Data processing errors

Examples:
- data input/keystroke errors
- imputation errors
- programming errors
- incomplete treatment of nonresponse

Dimension 4 Respondent bias/misrepresentation

Examples:
- falsification
- noncompliance
- altered behavior

Aspect II: Relevance

Dimension 5 Misapplication

Examples:
- using an unbiased sample to represent a biased set
- using data out of context (time, referent, or incentive structure)

Dimension 6 User bias/misrepresentation

Examples:
- sophistry
- selective perception

Figure 3. Classification schema for dimensions of credibility.

time (without adjustments for production rates). These data reported changes in consumption—they did not report conservation, but:

Consumption isn't consumption. The JPRS (Joint Petroleum Reporting System) was used as a source of information about petroleum demand and consumption (which are not necessarily the same thing), and yet what JPRS collected was withdrawals from available supply. These data reported changes in supply—they did not report consumption, but:

Supply isn't supply. The FPC (Federal Power Commission) Form 4 collected fuels information from power plants, fuels consumption, and fuels shortage data, but a power plant's endurance (i.e., its ability to continue to operate in the face of a cutoff in fuel supply) is not measured by dividing daily consumption data into average amount of fuel on hand. (Some of the "supply" on hand cannot be consumed.) These data reported changes in inventory—they did not report supply, and certainly not endurance (and yet they were routinely used to estimate capacity to endure a supply interruption).

When the government asks someone to answer a data-related question, the retort "Why?" is not necessarily an evasion, nor must it be a prelude to: "It's none of your business." It is simply an indication that the respondent knows that his or her answer depends on why the data are wanted. We have learned to recognize the credibility aspects of this dependency so well that we have overlooked the more fundamental possibility that the answer depends on what the data will be used for (i.e., it depends on what the question really is). All too frequently, the errors in an analysis are blamed on the data, when the errors actually stem from the assumption that data collected in one setting can be used to answer a question that occurs in another setting.

Albert Einstein frequently said, "The answers one gets depend upon the questions one asks" (Clarke, 1971). This may seem trivial, but it is a profound truth. Even at the simplest level at which one can formulate a question, there is an interaction between how the question is posed and how it is answered.

Some questions are silly. (If the average car gets 13 mpg, how big is a coal mine?) The fact that a grammatical statement has a question mark at the end is not an adequate reason to assume that there is an answer at all. Similarly, the fact that a communication is expressed in numbers is not an adequate reason to assume that it contains information.

Some questions that seem silly have reasonable answers. (If Ohio collected $18 million in general sales tax, what is its population?) The Bureau of Census uses general sales tax as an indicator for population size between census periods.

But every question constrains the answers. (If there is a glut of natural gas, will the price go down?)

Some questions aren't meant to be answered. (How do you do?)

And many questions have several different answers. (How long is the New Jersey shoreline?)

After struggling for some time with the implications of these kinds of questions and considering a variety of definitions of accuracy and validity, we came to the conclusion that each concept is contextual, and we realized that we needed to understand the forces that cause one to examine data for accuracy and validity. Therefore, borrowing from basic physics the notion of a *field*, referring to the host of observable phenomenona surrounding a subject, we developed a field theory of information, wherein we researched observable phenomena related to data sets (systems and models) and developed a notion of the field effects, associating each datum with a relevant field.

5.2 A Field Theory of Information*

The following paragraphs present a brief introduction to a field theory of information and argue that an appreciation of the theory will improve the quality of an analysis while reducing the amount of effort used in the search for unnecessary accuracy.

In the field theory, credibility is made up of many dimensions, all of which are important and necessary to data validation. But to illustrate the theory, either the dimensions have to be combined (e.g., accuracy vs. relevance, collection vs. use, honesty vs. dimensions that have no bearing on honesty), or the illustration has to be limited. For the purpose of this discussion, a two-dimensional representation will be used, with relevance representing two dimensions and accuracy representing four, as shown in Figure 3. (These two dimensions echo the concepts of meaningfulness and accuracy in the Data Validation Methodology.)

Using this representational approach, the field theory can be further illustrated by discussing both the structural nature and the dynamic nature of the field, keeping in mind that data or evidence may be classified as valid *with respect to a particular problem.* The structural nature of the theory refers, then, to the classification; the dynamic nature refers to the particular problem.

*For the classical example of a field theory applied to a setting involving human factors, see: Lewin, K., *A Dynamic Theory of Personality*, New York, McGraw Hill, 1935.

On the Structural Nature of the Field

On the structural side, it is to be noted that, in any given problem area, data or evidence may be classified as:

1. Possessing, or not, some adequate level of accuracy (as defined by the representational model), and
2. Bearing, or not, some reasonable level of relevance (again, as defined by the model).

Thus, all information is put into one of four categories for any given problem:

1. Data that are relevant but not accurate
2. Data that are both relevant *and* accurate
3. Data that are accurate, but not relevant
4. Data that are neither accurate nor relevant

These four categories are described below.

Relevant but not accurate. This category deals with data that are useful but not credible. It refers to data that are relevant to an analysis, but not persuasive to the users of the analysis. An example of this might be discounted testimony in a court of law, or data presented in political debates or in technical debates that have been politicized. This category may also refer to data so reliably biased that they can be corrected, in the way one might adjust for known constant errors of measurement. Or it may refer to data that simply seem incredible. (This is the projection of the user-bias dimension of credibility and includes such things as reports in 1491 that the earth was round.) An energy example of this category, for some users, might be any data that purport to show that nuclear power is necessary or reasonably safe.

Relevant and accurate. This category refers to credible data, but it is much more interesting than that simple characterization implies.

Most people dealing with data make the assumption that the data are both relevant and accurate; that is, they assume that the data are somehow relevant to the discussion and that they are to be believed. Our culture is reasonably adept at challenging the structure or the logic of a seemingly factual argument, but it is slow to question the veracity of the facts. This is especially true of numbers. People seem to confer upon numbers an honesty and precision that they would not dream of crediting to unsubstantiated text. However, it is also true that once people find cause to discount the meaning of numbers, they do so in an even more exaggerated fashion than they do with words. Most data are met with the assumption that they are both accurate and applicable, but once the data are seen as leading to an unwanted conclusion, they are generally rejected as inaccurate (i.e., incredible), with little or no thought given to the possibility that they might be accurate but not applicable. This is either the effect of a change along the relevance dimension or, more probably, the action of the user's bias; that is, the sixth dimension collapsed onto this two-dimensional space.

An energy example of this category is the data in DOE's Monthly Energy Report on gasoline sales. These data are both accurate and relevant (to gasoline supply and consumption analyses) and they are generally dealt with as though they were accurate and relevant. However, whenever these data lead to untenable conclusions (e.g., that there is "enough" gasoline when people are waiting in gasoline lines), then the data are suspected of being inaccurate (biased), when in fact they are accurate but not applicable. The data lost applicability because "sales" are not "consumption" and because inventories on hand at bulk terminals are not inventories on hand at retail stations. Under business as usual, these distinctions do not matter, but once a supply shortage is perceived and panic buying and hoarding begin, the data may retain their accuracy, but, because the situation has changed, the data have lost their relevance—to the question of supplies at retail stations. These now become data that belong in the "accurate but not relevant" area.

Accurate but not relevant. Ordinarily, one is not concerned about data that are credible but not useful. For most problems, this is the ocean of generally accepted knowledge that has nothing to do with the problem at hand. Thus, if one is studying energy, for example, virtually all of the "established" (i.e., credible) data from such fields as medicine, astronomy, or psychology are not applicable or only marginally relevant to the issues under study.

However, from the point of view of a field theory of information, data that are credible (assumed to be without bias) but not (yet) known to be relevant are the most interesting kind of data. Observing the relevance of data from this category is the source of much creativity, and failing to appreciate their relevance is the source of much error. Data available at any given time represent efforts to answer questions posed in the past. There is always a time lag between the posing of the question and the availability of the answer. Therefore, the challenges facing those who would use off-the-shelf data are to (1) figure out what to do with new questions, (2) develop improved ways of posing questions so that the answers are more robust (i.e., more useful for other questions), and (3) develop better ways of anticipating future issues and getting ready to deal with them.

In the vast majority of cases where a problem has been solved with a stroke of insight, a key to that solution will be found to have been the use of one or more information elements that are credible (accurate and relevant), but that had previously been seen as not applicable. Indeed, one of the standard definitions of creativity is the perception of new relationships between theretofore unrelated bits of information (Woodworth, 1938). In the context of this discussion, it is the recognition that information that had been thought to be *not* applicable, was, in fact, applicable.

On the other hand, in the vast majority of cases where a problem has bogged down into dispute and diatribe, a key to the collapse of objectivity will be found to have been the inappropriate adoption of one or more elements of information that are prima facie credible, but, for one reason or another, are not applicable. Indeed, our work in energy information validation has shown that in virtually every case where an energy analysis was said to be vulnerable on the grounds of inaccurate data, it was subsequently found to have used reasonably accurate data in an inappropriate (nonrelevant) manner (Taylor, 1980; Taylor, 1981).

Neither accurate nor relevant. Material that is neither accurate nor relevant is not generally called by any of the word names that connote data. Examples of material that is generally thought by the mainstream of science to be neither accurate nor relevant are astrological assessments, UFO sightings, drug-induced visions, and children's representations of their imaginary adventures. None of this material is accepted as objective facts in the mainstream of discourse, though all of it is properly attended to in its own setting.

However, this category of data is a convenient vehicle for noticing the determinants of relevance. In fact, it is the setting itself that defines applicability when the question arises. Fossil records are applicable in any setting, but geological interpretations of those records are not necessarily applicable in a fundamentalist discussion of emanation versus evolution. Similarly, accuracy and relevance converge (on credibility) whenever there is a superordinate context that is exempt from critical appraisal. If there is something that must be taken on faith, then data that conflict directly or indirectly, or that are seen as being in potential conflict (not credible), are either or both not relevant or not accurate.

On the Dynamic Nature of the Field

On the dynamic side of field theory, it is necessary to note that the field changes with time as well as with the problem. Thus, an analyst who was dealing with the "right" set of information for Problem Y at Time 1 might find that that same set of information was no longer appropriate for Problem Y at Time 2.

The dynamic nature of the field is such, therefore, that from moment to moment, as the flux and flow of information changes, astute problem solvers (information users) stay alert to the possibility that neither the foundation for their analyses (what was known when they started) nor the implications of their findings are static.

The time frames involved need not be very long for this effect to be apparent. Most information analysts are familiar with problem areas where five years later there are new elements in the set of materials that are considered relevant. Energy is a case in point.

Numbers are no more infallible, and not necessarily any more precise, than are words. If information (in a communications sense of the word—Shannon, 1949) is a reduction in uncertainty, and if conflict (in a psychoanalytic sense of the word—Freud, 1952) is an increase in uncertainty, then when one's analysis is not persuasive, the "data" are probably not information.

6. SUMMARY

The conclusions of our study can be summed up by the following definition: "Validation is the process of determining how strong and authoritative a database or model is for its intended uses." This report demonstrates the truth of this definition by reviewing the origins, methods, and findings of an energy-data quality-assurance program conducted by Oak Ridge National Laboratory from 1978 to 1981. The thrust of the chapter has been on the broad implications of this work, rather than on narrow details that may apply only to the context in which the work was carried out. The conclusions of the Chapter are that data are generally viewed too literally. Most people (including most analysts) take numbers (data) too uncritically at first sight, then reject them too totally when they exhibit some inaccuracies.

Significant dimensions of the research problem that ORNL addressed and the basic characteristics of the approach were illustrated with examples drawn from the work that was done. The illustrations were designed to demonstrate (1) the foundation of findings that we used to support our conclusions, and (2) the breadth of applicability of those findings.

The background section of the chapter described the environment of skepticism and mistrust that surrounded energy data issues in the middle to late 1970s. The interaction of this aura with ORNL's research management philosophy was pointed to as the basic impetus behind a systems approach to addressing energy data validation. The chapter then described ORNL's problem-solving approach, with special emphasis on the characteristics that contributed most to its success. Considerable attention was given to both the generic research strategy and the specific application of that strategy to the evolution of the problem at hand.

The results reported fall into three categories. Woven throughout the first major sections are observations on what was learned or demonstrated about the technical aspects of research management in an evolving, multidisciplinary, controversial, quasi-centrally controlled project. The second set of results is reported as three major conclusions that are drawn from, but do not apply exclusively to, energy data:

While every data-collection system has a host of opportunities for random error, the systematic errors in large-scale data-collection systems are much more significant than are the random errors.

While conscious bias or misrepresentation are inevitably possible, they are likely to cause massive distortion of data under only one set of circumstances (when there is a low-risk incentive to distort).

The single most significant source of error in data analysis is misapplication of data that would be reasonably accurate in the right context.

The final category of results was characterized as a general field theory of the credibility aspects of data. The argument is made that data credibility ought to be viewed as a multidimensional aspect of data, wherein, on at least one set of dimensions, data can belong to a field in which they can have both accuracy and (changing) relevance characteristics. The proposition is put forth that since both characteristics are essential for a valid analysis, and since accuracy is the customary direction in which analysts look when they need to explain difficulties in their analyses, it follows that lack of relevance is the most likely source of data-credibility difficulties.

REFERENCES

Ackoff, R. L., *The Design of Social Research*, Chicago, University of Chicago Press, 1953.

Brais, G. F., et al., *Data Validation of the Voluntary Business Energy Conservation System and the Industrial Energy Efficiency Improvement Program*, Vienna, Va., Evaluation Research Corp., May 1974,

Chapanis, A., W. R. Garner, and C. T. Morgan, *Applied Experimental Psychology*, New York, John Wiley and Sons, 1949.

Churchman, C. W., *Methods of Inquiry*, St. Louis, Educational Publishers, 1950.

Churchman, C. W., R. L. Ackoff, and E. L. Arnoff, *Introduction to Operations Research*, New York, John Wiley and Sons, 1957.

Clarke, R. W., *Einstein: The Life and Times*, New York, The World Publishing Co., 1971.

Conklin, Bert, Memorandum for T. C. Reed, Executive Office of the President, Washington, D.C., Aug. 12, 1977.

Evaluation Research Corporation, *Validation of the Capacity of Petroleum Refineries System*, Oak Ridge National Laboratory, ORNL/Sub-7502/1, March 1979.

Freud, Sigmund, *The Psychopathology of Everyday Life*, New York, NAL, 1952.

General Accounting Office, *Actions Needed to Improve Federal Efforts in Collecting, Analyzing, and Reporting Energy Data*, B-178205, Washington, D.C., Feb. 6, 1974.

General Accounting Office, *The Energy Information Act, S.* 1864, Mar. 9, 1976.

General Accounting Office, *Improvements Still Needed in Federal Energy Data Collection, Analysis and Reporting*, OSP-76-21; B-178205, Washington, D.C., June 15, 1976.

Hertz, D. B., *Theory and Practice of Industrial Research*, New York, McGraw Hill, 1950.

Koopman, D. B., *Federal Energy Management Program*, Vienna, Va, Evaluation Research Corporation, August 1978.

Lancaster, M. A., *Error Profile of the Motor Fuel Taxation Data Used to Establish and Monitor State Emergency Conservation Targets*, Falls Church, Va., Transportation and Economic Research Associates, Inc., February 1971.

Liepins, G. E., *Data Analytic Tools in Support of Data Validation* (Draft Report), Oak Ridge National Laboratory, June 1980.

Murray, H. A., et al., *Assessment of Men*, New York, Rinehart and Company, 1947.

MAXIMA Corporation, *Storage Structures and Access Methods for Statistical Data: A Review and Analysis of Performance*, Oak Ridge National Laboratory, December 1980.

Oak Ridge Associated Universities, *Validation of the Natural Gas Curtailments System*, Oak Ridge National Laboratory, August 1979.

Oak Ridge National Laboratory, *Validation of the Monthly Power Plant Report System*, DOE/EIA-0254, October 1980.

Professional Audit Review Team, *Report to the President and Congress: Activities of the Energy Information Administration, Department of Energy*, Washington, D.C., May 7, 1979, pp. 12–13, 18–20.

Shannon, C., The Mathematical Theory of Communication, *Bell System Technical Journal*, July and October 1948. Reprinted in book form, Urbana, Ill., University of Illinois Press, 1949.

Shewmaker, John, Office of Energy Information Validation, *Briefing to the New Administrator*, April 1981, Unpublished.

Taylor, D. A., *A Preliminary Assessment of the Accuracy of Energy Information Data Related to Volumes of Petroleum*, Washington, D.C., Department of Energy, Energy Information Administration, Office of Energy Information Validation, May 1980.

Taylor, D. A., *An Assessment of the Accuracy of Principal Data Series of the Energy Information Administration*, Washington, D.C., Department of Energy, Energy Information Administration, Office of Energy Information Validation, June 19, 1981.

Transportation and Economic Research Associates, *Data Validation Study of the Prime Suppliers Monthly Report*, Oak Ridge National Laboratory, April 1980.

Transportation and Economic Research Associates, *Evaluation of Federal Highway Administration Form MF-33G State Gasoline Consumption Data* (Draft Final Report), July 1981.

Tukey, J. W., *Exploratory Data Analysis*, Addison-Wesley Publishing Co., Reading, Mass., 1977.

Underwood, B. J., *Design of Experiments*, New York, Appleton Century Crofts, 1956.

Validation of Information Requirements for Commercial Sector Energy Consumption, Subcontract Report, BDM, August 1980.

Von Neuman, J., and O. Morgenstern, *Theory of Games and Economic Behavior*, Princeton, N.J., Princeton University Press, 1947.

Wiener, N., *Cybernetics*, New York, John Wiley and Sons, 1948.

Woodworth, R. S., *Experimental Psychology*, New York, Holt, 1938.

10

Editing and Imputation of Multivariate Data: Issues and New Approaches

RODERICK J. LITTLE University of California at Los Angeles, Los Angeles, California

1. INTRODUCTION

In recent years, survey research has turned increasingly from the issues of probability sampling and sampling error to nonsampling components of survey error. Part of this change may be attributed to the notable successes of research into scientific sampling methods. These methods, in my view, are among the most useful tools created by statisticians (and I speak as a modeler unconvinced by randomization as a mode of inference!). When an efficient sample design in a large survey leads to modest sampling errors, the relative importance of nonsampling error is increased. The significance of the topic of this conference is clear, since data quality control concerns methods for the reduction of nonsampling errors in the design, conduct, and analysis of surveys.

Section 2 contains some general remarks on the design of data-collection efforts to limit the impact of nonsampling errors. The remainder of the chapter concerns analysis. Section 3 discusses the role of statistical models in refining editing and nonresponse adjustments. Section 4 discusses some univariate and multivariate models for outlier detection, with emphasis on extensions of the ratio model. Section 5 discusses data analysis in the presence of outliers. Finally, Section 6 reviews recent modeling literature on missing data, with an emphasis on Rubin's multiple imputation methods.

2. PLANNING FOR MISSING DATA

Very few large-scale surveys yield data that are entirely complete and free from errors. Thus, methods for limiting the impact of nonsampling error and nonresponse should be considered in the design of the survey. These methods may be directed at limiting the extent of the problems or at providing better adjustments at the analysis stage.

As a general principle, imputation of a missing datum is a poor substitute for having the true value, particularly when the missing-data mechanism is not under the control of the surveyor. The most direct and satisfying approach to nonresponse is to make it as rare as possible. Field strategies for limiting *unit* nonresponse seem particularly important, since analytical methods of adjusting for unit nonresponse are often limited by lack of information on the nonrespondents. Madow, Nisselson, and Olkin (1983, Part 1 Chapter 4) list 14 practical steps for limiting nonresponse in the data-collection phase. Suggestions include limiting respondent burden, interviewer training to limit refusals, provision of evidence of confidentiality, intensive callback strategies, reminder postcards for mail surveys, and attempts to convert refusals, perhaps using a shorter questionnaire that conveys key information.

When nonresponse occurs, a useful broad objective is to obtain as much relevant covariate information about nonrespondents as possible, to help predict missing outcomes. The covariates should be good predictors of missing variables. Also, unless the predictive distribution relating the missing variables to the covariates is known, the covariates also need to be measured for respondents, so that the distribution can be estimated from the data.

The point can be illustrated using the simple data patterns in Figure 1. Here, X denotes an incomplete variable and Z denotes a variable (or set of variables) measured for nonrespondents to X. In Case (A), X and Z are both recorded for respondents. These data can be used to estimate the conditional distribution of X given Z for imputing the missing values of X. However the correlation between X and Z is weak, so Z is a relatively poor predictor of X. A common instance of Case (A) is unit nonresponse, where the nonrespondents are missing all survey variables and Z is limited to survey design variables such as geographic location.

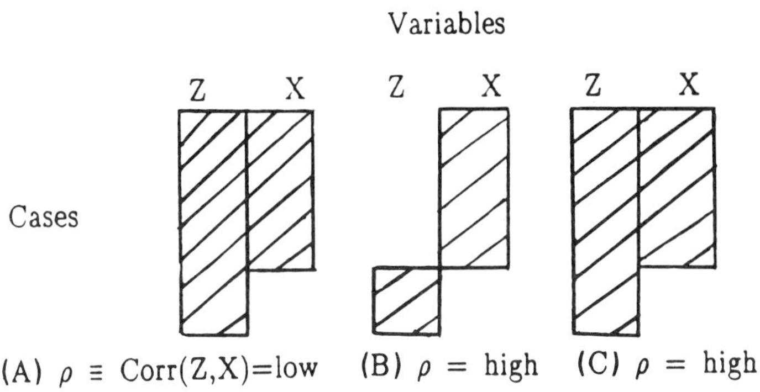

Figure 1. Three simple missing data patterns.

In Case (B), X and Z are highly correlated, but Z is not recorded for respondents to X, so the conditional distribution of Z given X cannot be estimated from the data. A common example occurs when Z is a proxy for X from another source; for example, X is income and Z is reported income from a spouse or other relative. Then Z is a highly correlated but imperfect predictor of X, but we obtain Case (B) unless we make a point of collecting both X and Z for respondents. We are led to assume that $X = Z$, acting as if proxy information is perfect.

Case (C) illustrates the best situation, where X and Z are highly correlated and observed together for a subsample of cases. Unfortunately, Case (C) is less common than are Cases (A) and (B) in practice, although it can arise in item nonresponse problems. Note that if Z represents proxy variables, then collecting proxy information Z for respondents to obtain data as in Case (C) may yield excellent data for X. The regression of X on Z provides information for checking and modifying the default prediction $X = Z$ and for estimating the error from replacing true by proxy information. Examples of data of this form are the Current Population Survey/Internal Revenue Service (CPS/IRS) match files, where proxy income information for CPS nonrespondents is obtained by a match to IRS records (Welniak and Coder, 1980). However, these files are used for research purposes and are not integrated into the design and analysis of the CPS. The match files attempt to obtain IRS data for all respondents, but a carefully designed respondent subsample (possibly matching on nonrespondent characteristics) may suffice or even be preferable. There seems to be a need for design and analysis work in this area.

In the context of data editing, Hinkins and Scheuren (1986) discuss a useful design option that leads to data in the form of Figure 1, Case (C). Since detailed editing of every questionnaire is expensive, intensive editing is limited to a random subsample of the data, possibly stratified on variables thought to be predictive of edit changes. The whole sample is subject to a more cursory edit, perhaps limited to automatic machine corrections. The result is data of the form of Figure 1, Case (C), with X = fully edited data and Z = partially edited data. Here, the missing-data mechanism is under control of the sampler (that is, it is *ignorable* in the sense discussed in Little and Rubin, 1987), so analysis of the data is less subject to assumptions than in the earlier nonresponse setting.

Questionnaire design is another area where the effects of nonresponse can be limited at the data-collection phase. In particular, questions should be carefully designed to distinguish among nonresponse, zero outcome, and not in universe. The distinction between nonresponses and zeros seems particularly important, since results can be strongly affected by which of these alternatives is considered correct. If the distinction is unclear, the problem seems hard to fix at the analysis stage, requiring either strong assumptions about the frequency of zeros among nonrespondents or a follow-up sample of the nonrespondents to determine their true status.

3. EDITING AND NONRESPONSE ADJUSTMENT: THE ROLE OF MODELS

I now turn to editing and nonresponse adjustments after the data have been collected. A distinction can be drawn between methods based on a priori information available before data are collected and methods based on a posteriori analysis of the data. The former methods include identity constraints, such as the requirement that components of a mixture sum to one, and inequality constraints for ratios determined by prior knowledge about the variables in the population. The latter methods include outlier-detection methods based on sample distributions of variables measured in the survey or on model-based nonresponse adjustments. Both methodologies have their uses. A priori methods have the operational advantage that they can be specified in advance before the data arrive. This may allow speedy detection of edit violations that can be resolved in the field, particularly if the edit checks are incorporated in computerized data-entry systems. Problems found at the analysis stage may appear too late to be resolvable at the data source. However, statisticians are quick to criticize investigators who massage their data to conform to preconceptions. Editing and imputation based on a priori information have a somewhat similar flavor.

Standard procedures for catching gross errors, such as range checks based on marginals and structure checks to see that questionnaire logic is followed, are clearly useful and appropriate. Beyond this, what can be done to improve practical methods for editing and imputation? One strategy is to (a) relate methods to underlying models for the population, (b) refine the implied models with respect to a priori plausibility and consonance with the data, and (c) adopt procedures implied by the refined models. I favor this modeling approach, not because I believe that models are true, but because model criticism and refinement serves as a bridge to better analysis. Many methods currently used for editing and imputation are intuitive and ad hoc; while intuition certainly has a role, it has limits when dealing with complex problems involving incomplete multivariate data. Models provide an underlying principle for generating and assessing methods.

A conceptual advantage of the modeling approach is that it places edit and nonresponse adjustments in the proper context: the statistical analysis of the data. I think it is important to see editing and imputation not as ends in themselves, but as aspects of data analysis. The emphasis should be on *appropriate estimation and inference from the adjusted data, not on optimal adjustments for particular values in the data.*

To show that this perspective matters, consider imputing the mean of a continuous variable. An optimal imputation of a missing value (in the mean-squared-error sense) is its conditional mean given observed covariates and edit constraints. However, mean imputation has severe limitations when the filled-in data are analyzed: Distributions are distorted, and standard errors are underestimated. From the analytical perspective, random draws from the predictive distribution are often preferable, despite the fact that as predictions of the individual missing values they are relatively poor. (See Rubin, 1987, and Little, 1988a, for more discussion.)

As another example, consider imputing a binary attribute Y (1 if present, 0 if absent) with no covariate information. Half the cases are observed, half the cases are missing, and 10 percent of observed cases have the attribute. It seems reasonable to restrict imputes to valid codes (0 or 1). Clearly, given this restriction and the available data and assuming that nonrespondents are like respondents, the best prediction of the missing values is $\hat{Y} = 0$; this choice will minimize misclassification errors. However, if this "optimal" impute is applied to all the nonrespondents, the resulting estimate of the proportion with the attribute is 5 percent, clearly too low. A procedure that imputes 0 with probability 0.9 and 1 with probability 0.1 provides a better estimate of the proportion, although it yields an inferior misclassification rate. (Of course, analysis of the complete cases may be preferable to any form of imputation here, since it does not overstate sample size; a more realistic example would introduce covariates, but the main point would still apply).

A final example concerns the question of whether imputations should be forced to satisfy all a priori edit constraints. This requirement seems obvious and is not inconsistent with the modeling perspective, in that an appropriate model should take into account such constraints and, thus, produce imputations that satisfy them. However, from an analytical perspective, it is possible to achieve valid inferences about aggregate quantities from data with individual imputes that violate edit constraints. Thus, I can imagine adopting a model that violates edit constraints in the interests of simplicity, *if the loss of precision is tolerable and the resulting inferences are defensible.* The satisfaction of edit constraints is not the predominant concern, as tends to be the case when the emphasis is on the propriety of the individual imputes. I now develop these ideas with reference to some specific problems.

4. OUTLIER DETECTION

4.1 Methods for Ratios

Many surveys employ edits involving ratios of positive-amount variables. A priori edits often take the form of ratio-inequality constraints:

$$l \leqslant Y/X \leqslant u \tag{1}$$

where Y/X is the ratio and l and u represent bounds determined by subject-matter specialists, presumably based on prior experience and data. For example, many variables in establishment surveys are characterized by heavy right-tailed distributions, due to marked disparities in size. Ratios tend to cancel these size effects and yield more powerful variables for detecting outliers.

Ratio-inequality constraints can be effective tools and play an important role in some fine editing systems (Sande, 1979; Greenberg, 1982). However, the choice of upper and lower limits may place too much reliance on the present looking like the past; also, the input of subject-matter specialists may be expensive. Fellegi (1975) argues that, in many cases, ratio-inequality constraints might be replaced by automatic procedures based on a posteriori methods of multivariate outlier detection. I consider multivariate methods in Section 4.2, but first discuss outlier detection based on the univariate distribution of a single ratio $Z = Y/X$.

A simple approach is to treat as outliers cases i with ratios that fail the constraint

$$\bar{z} - ks \leqslant z_i \leqslant \bar{z} + ks \tag{2}$$

where $\bar{z}$ is the sample mean ratio, s is the sample standard deviation of the ratios, and k is a tuning constant chosen subjectively (for example, $k = 3$). More robust estimators of location and scale may be preferable to avoid masking effects. Hidiroglou and Berthelot (1986) point out a more fundamental problem with this approach; namely, that it fails to take into account *heteroskedasticity;* intuitively and as an empirical fact, the variance of Y/X tends to decrease with increasing X. As a result, (2) will be relatively insensitive to outlier ratios with large values of X. Hidiroglou and Berthelot also note that a natural estimate of location here is not the average ratio but the ratio of averages:

$$\tilde{z} = \sum_{i=1}^{n} y_i \bigg/ \sum_{i=1}^{n} x_i \tag{3}$$

where i indexes cases.

From the modeling perspective, (2) assesses lack of fit from the normal model

$$z_i \sim {}_{\text{ind}}N(\mu, \sigma^2) \tag{4}$$

in that $r_i = (z_i - z)/s$ are standardized residuals from the least-squares fit of this model, and (2) flags values of r_i that are larger than k. I propose to detect outliers using standardized residuals from refinements of (4). To model heteroskedasticity, (4) is replaced by

$$z_i \sim {}_{\text{ind}}N(\mu, \sigma^2/x_i^\lambda) \tag{5}$$

for some positive λ. An appropriate standardized residual for this model is

$$r_i(\lambda) = x_i^{\lambda/2}[z_i - \hat{\mu}]/\hat{\sigma}$$

where $\hat{\mu}$ and $\hat{\sigma}$ are the standard weighted-least-squares estimates of μ and σ. The corresponding modification of (2) is to flag cases i with residuals

$r_i(\lambda)$ that are greater than k or, equivalently, for which

$$\hat{\mu} - k\hat{\sigma} x_i^{-\lambda/2} \leqslant z_i \leqslant \hat{\mu} + k\hat{\sigma} x_i^{-\lambda/2} \tag{6}$$

Without looking at data, a natural choice of λ is one, since then (5) is the ratio model and $\hat{\mu} = \tilde{z}$, as in (3). However, λ need not be fixed a priori, if current or historical data contain information about the appropriate choice. For example, λ might be assessed as minus the slope of a plot of log $r_i^2(0)$ against log x_i, ignoring outliers.

This approach can be compared with the procedure for dealing with heteroskedasticity suggested in Section 2.1.2 of Hidiroglou and Berthelot (1986), which lacks the organizing principle of an underlying model and requires three tuning constants to be specified by the analyst. Note that (6) has only two tuning constants (k and λ), and one of these can be estimated from the data.

We can do more. Note that (5) implies that the mean of Z is the same for all values of X. This assumption may be incorrect and can be readily tested on the data by computing the weighted regression of Z on X. More generally, we might regress Z on a set of q regressors $V_1, \ldots, V_q$, perhaps including X, thought to be predictive of Z; that is, fit the model

$$z_i \sim_{\text{ind}} N(\beta_0 + \beta_1 v_{i1} + \cdots + \beta_q v_{iq}, \sigma^2/x_i^\lambda) \tag{7}$$

For $\lambda = 1$, I like to call this the extended-ratio model (Little 1988b). Again, large standardized residuals from the model are flagged as outliers.

Refinements of this approach (a) screen outliers when estimating the regression or adopt a robust regression method, and (b) studentize the residuals if sample sizes are small. However, note that the approach requires only readily available weighted regression software. Most programs provide a weighting option, although care is needed to ensure that they provide correctly weighted residuals.

4.2 Multivariate Outlier Detection Methods

Ratio edits confine information to pairs of variables; checks based on the joint distribution of a set of variables may be more revealing. For p quantitative variables $Y_1, \ldots, Y_p$, a natural approach is to compute the distance from the mean normed by the covariance matrix:

$$d_i^2 = (\mathbf{y}_i - \bar{\mathbf{y}})^T S^{-1} (\mathbf{y}_i - \bar{\mathbf{y}})$$

Under normal assumptions,

$$F_i = \frac{(n-p)n}{p(n-1)(n+1)} d_i^2$$

is F distributed with p and $n - p$ degrees of freedom (Anderson, 1958; Hawkins, 1974). Thus, the d_i^2 can be transformed to standard normal deviates and plotted against normal order statistics to detect outliers (Gnanadesikan, 1977). A good transformation to approximate normality is

$$z_i = \frac{\left[\left(1 - \frac{1}{2v_2}\right)F_i\right]^{1/4} - \left[1 - \frac{1}{2v_1}\right]^{1/4}}{\left[\left(\frac{F_1}{8v_2}\right)^{1/2} + \left(\frac{1}{8v_1}\right)^{1/2}\right]^{1/2}} \tag{8}$$

where $v_1 = p$ and $v_2 = (n - p)$ are the numerator and denominator degrees of freedom. This expression is obtained by transforming numerator and denominator chi-squared deviates in the F-statistic to normality using the fourth-root transformation suggested by Hawkins and Wixley (1986), and then applying Geary's approximation for the ratio of two normals (see Kendall and Stuart, 1969, Ex. 11.12). Little and Smith (1987) develop this approach by protecting the estimates of the mean and covariance matrix from outliers and allowing some values in the data matrix to be missing. The estimates $\bar{\mathbf{y}}$ and $\mathbf{S}$ are replaced by estimates from a robust modification of the EM algorithm (Dempster, Laird and Rubin, 1977) for incomplete multivariate normal data, and d_i^2 is restricted to the observed variables in each case.

Hawkins (1974) notes that outlier tests based on d_i^2 are not optimal for testing for outliers in particular variables. Thus, the method might be supplemented by inspection of the individual marginals of the variables. Another approach is to restrict attention to linear combinations of variables that have low variance in the sample, since for these variables, outliers will be particularly noticeable. These combinations may be suggested by prior information; for example, if a set of components that sum to one are measured with small independent errors, then the distribution of the sum should be centered at one and have a small variance; the sum is then a natural linear combination to examine for outliers.

Empirically, variables with low variance can be found from a principal component analysis of $\mathbf{S}$, as eigenvectors corresponding to the small eigenvalues. Thus, for outlier detection, one might confine attention to eigenvectors corresponding to low eigenvalues, an approach that has the additional advantage that the individual eigenvectors are uncorrelated. The method will work best if $\mathbf{S}$ is protected from outliers by a robust estimation procedure or a preliminary screening to eliminate obvious outliers. This approach is considered by Hawkins (1974) and Fellegi

(1975). Difficulties include the choice of how many eigenvectors to include and the fact that a simple distribution theory is lacking even under multivariate normal assumptions.

Covariates $\mathbf{V} = (V_1, \ldots, V_q)$ can be introduced here by assessing lack of fit in the residual space. Let $\hat{\mathbf{B}}$ and $\mathbf{S}$ be the least-squares estimates of $\mathbf{B}$ and $\boldsymbol{\Sigma}$ from the multivariate regression model

$$\mathbf{y}_i \sim {}_{\text{ind}}N_p(\mathbf{Bv}_i, \boldsymbol{\Sigma}) \tag{9}$$

where N_p denotes the p-variate normal distribution. Then, for case i, we have

$$d_i^2 = (\mathbf{y}_i - \hat{\mathbf{B}}\mathbf{v}_i)^T S^{-1}(\mathbf{y}_i - \hat{\mathbf{B}}\mathbf{v}_i) \tag{10}$$

$$F_i = \frac{(n - qp)n}{p(n-1)(n+1)} d_i^2 \tag{11}$$

and z_i is given by (8) with $v_1 = p$, $v_2 = n - qp$.

The assumption of multivariate normality of the *Y*s may be inappropriate for skewed quantitative variables. If the variables are strictly positive, then transforming to the log scale (the first rule of data analysis!) may be useful; Little and Smith (1987) argue for the log transformation, since it has the effect of linearizing ratio-inequality edits. Another approach is to extend the heteroskedastic-ratio model of the previous section to a multivariate outcome. Let x_i denote a nonnegative scalar variable representing some measure of the size of a unit, and let $\mathbf{z}_i = \mathbf{y}_i/x_i$ denote the vector of ratios $(y_{i1}/x_i, \ldots, y_{ip}/x_i)$. Consider the model

$$\mathbf{z}_i \sim {}_{\text{ind}}N_p(\mathbf{Bv}_i, \boldsymbol{\Sigma}/x_i^{\lambda}) \tag{12}$$

The generalized residual from this model is

$$d_i^2(\lambda) = x_i^{\lambda}(\mathbf{z}_i - \hat{\mathbf{B}}\mathbf{v}_i)\hat{\boldsymbol{\Sigma}}^{-1}(\mathbf{z}_i - \hat{\mathbf{B}}\mathbf{v}_i)^T \tag{13}$$

where $\hat{\mathbf{B}}$ is the weighted-least-squares estimate of $\mathbf{B}$ and $\hat{\boldsymbol{\Sigma}}$ is the residual covariance matrix. Replacing d_i^2 in (11) by $d_i^2(\lambda)$ yields an F-statistic, which can be transformed to normality and plotted as before. As in the scalar model of the previous section, one is a natural choice for the parameter λ, yielding a *multivariate extended-ratio model.* Better choices of λ may be suggested as minus the slope of a plot of $\log d_i^2(0)$ against $\log x_i$. Computations remain in the realm of weighted least squares and existing software.

Two advantages of modeling ratios ($\mathbf{Y}/X$) rather than logarithms ($\log \mathbf{Y}$) are worth mentioning. First, zeros in $\mathbf{Y}$ or X are not a serious problem; cases with $X = 0$ are given a weight of zero, and cases with components of $\mathbf{Y}$ set to zero do not unduly influence the analysis. For analysis of log $\mathbf{Y}$, cases with zero $\mathbf{Y}$ components need to be excluded or the analysis modified by adding arbitrary constants to $\mathbf{Y}$. Second, imputation of missing or edited components of $\mathbf{Y}$ is simpler for the ratio model than for the log model, since the ratios are linear functions of $\mathbf{Y}$. A related issue on nonlinear transformations arises in Section 5.2.

A limitation of (12) is that negative values of $\mathbf{z}$ are not precluded; a better model would replace the normal distribution for $\mathbf{z}$ by a distribution with support on the positive real line (see Little, 1988b, for more discussion of the univariate case). However, replacing negative predictions from (12) by zero may be adequate for many applications.

EXAMPLE 1 We illustrate multivariate methods on a data set collected to test diagnostic procedures for detecting and monitoring side effects among patients treated by antipsychotic drugs; for example, for depression (May, Lee, and Bacon, 1983). The procedure measures tremor in different parts of the body as electrical activity in eight frequency bands, ranging from 0.125 to 32 Hz. This analysis focuses on measurements of tremor in the resting hand for 207 subjects, 129 controls and 78 depressed patients diagnosed as suffering from tardive diskinesia (TD). The variables consist of the eight activity measures $X_1, \ldots, X_8$ and a grouping variable V indicating control ($V = 1$) or TD ($V = 2$). Variations in the measurement protocol led to missing data. Specifically, X_1 and X_2 were missing on 5 controls, and X_1 and X_8 were missing on 94 controls.

The raw X measurements exhibited heavy right tails, so a transformation was indicated. One possibility is to apply the above multivariate-ratio model, using total energy ΣX_j as X, and removing one of the X variables to avoid the constraint that the resulting ratios sum to one. After preliminary analysis, it was decided to work with log-transformed data instead; specifically, $Y_j = \log(X_j + 0.19)$, where the constant 0.19 provided symmetric marginal distributions and also had some physical significance for the measurement instrument. The analysis was based on the multivariate regression model (9) for $\mathbf{Y}$, with a single covariate V consisting of the binary group indicator.

Figure 2 shows a normal plot of transformed distances (8) from this model. Here, the complete-data analysis based on (8), (10), and (11) needs to be modified to allow for missing values. In particular, d_i^2 is restricted to variables observed in case i, and $\hat{\mathbf{B}}$ and $\mathbf{S}$ are replaced by maximum-likelihood (ML) estimates of $\mathbf{B}$ and Σ computed from the normal EM algorithm. The small-sample correction from d_i^2 to F_i is computed using (11), with p replaced by the number of Y-variables observed for case i. The correction is approximate, since F_i is not exactly F-distributed when the data are incomplete.

From Figure 2, we see that the observed line deviates considerably from the line with unit slope expected under normality and, in particular, the last six cases are larger than the expected line given normal data. These cases can be checked for errors and possibly downweighted in subsequent analysis (see Section 5.3).

4.3 Localization of Outliers

When a case is flagged as an outlier by multivariate edit checks, it is not necessarily clear which field or fields are in error. Fellegi and Holt (1976) provide a comprehensive discussion of this topic for discrete variables. They propose changing the set of fields that cure edit violations with a minimal number of changes to the record and show how the set and the changes can be determined. For continuous variables, Little and Smith (1987) propose a stepwise procedure for successively removing values from an outlying record that yield the greatest reduction of the distance d_i^2, until an acceptable value of d_i^2 is obtained. This approach can be contrasted with localization methods based on ratio-inequality edits (Sande, 1979; Greenberg, 1982), which have a deterministic rather than a statistical flavor.

Methods for mixtures of continuous and categorical variables seem to have received little attention. One approach would be to apply the Fellegi and Holt methods to the categorical variables and then compute distances as generalized residuals from a multivariate analysis-of-variance model of the continuous variables on the categorical variables. Parameter estimation for this approach with incomplete data can be achieved by the ML methods of Little and Schluchter (1985) or, more simply, by multivariate analysis of variance applied to a subset of complete cases.

5. TREATMENT OF OUTLIERS

5.1 Outliers: True or False?

What can we do about outlying values? Where possible, the source of the information should be reviewed and errors corrected. It seems important to distinguish conceptually between outliers that are truly part of the population of interest (say, *true* outliers) and outliers that are aberrations

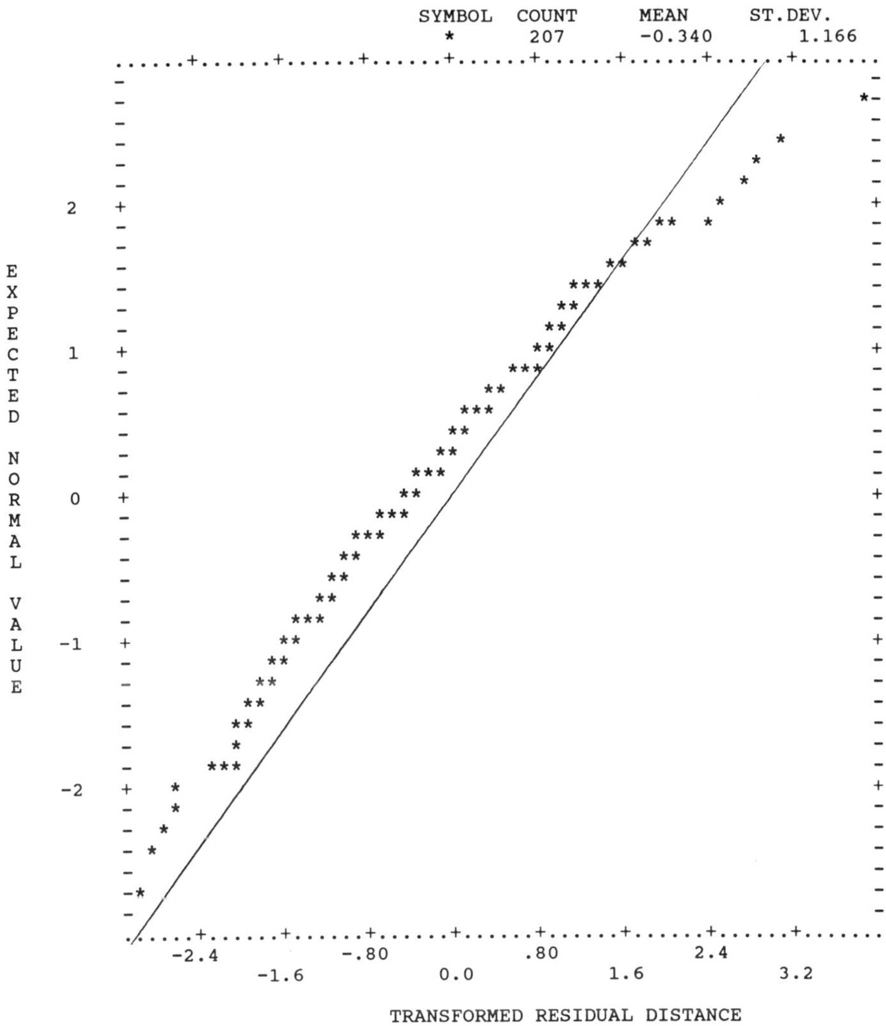

Figure 2. Normal plot of transformed residual distances, resting tremor data, normal model.

(say, *false* outliers); in practice, this distinction is often unknown or blurred.

Three strategies for analyzing data with outliers can be distinguished: (a) *do nothing*; that is, analyze the data without special treatment of the outliers; (b) *deletion*; that is, delete the outliers and analyze the imcomplete data, perhaps imputing for the deleted values; and (c) *accommodation*; that is, leave the outliers in the data set and use a robust analysis that limits their influence. For true outliers, (a) and (c) seem to me likely alternatives; (b) should be applied only with considerable circumspection. For false outliers, we might consider deletion or accommodation; (a) does not seem a sensible option. I now develop these distinctions with some examples.

5.2 Finite Population Inference in the Presence of True Outliers

Samples from populations with a small number of very large values create interesting analytical problems that deserve further study, in my view. Robust estimation methods may be of limited use, since most methods are developed for long-tailed symmetric distributions; whereas, in practice, distributions are often highly skewed, with long right tails. Transformations to symmetry, such as a logarithmic or power transformation, may help, particularly when inference about parameters on the transformed scale is acceptable. However, finite population quantities on the raw scale are often of primary interest, and transforming back to the raw scale requires care. For example, suppose that log (income) appears symmetrically distributed, but interest concerns the mean or total income in a population, not the mean or total of log (income). The sample mean $\bar{y}$ (or a robust modification) may be a good estimate of mean log (income), but exp ($\bar{y}$) is a biased estimate of mean income. The corrected estimate exp ($\bar{y} + s^2/2$) is appropriate if log (y) is normal, but the correction is highly sensitive to the normality assumption.

Rubin (1983a) provides an illuminating illustration of this point. The problem is to estimate the mean population size $\bar{Y}$ from a simple random sample of 100 of the 804 municipalities in New York State. The standard frequentist confidence interval based on the sample mean is highly erratic, and the assumption of normality based on Central Limit Theorem is not justified, since the distribution is highly skewed. In an attempt to model the observed distribution, Rubin applied a Bayesian model with Box-Cox transformations to normality. Two relatively well-supported values for the power transformation were $\varphi = -1/8$ (posterior probability 0.137) and $\varphi = -1/4$ (posterior probability 0.863). These values of φ gave 95 percent posterior intervals for $\bar{Y}$ of $(1.0 \times 10^4, 2.5 \times 10^5)$ and $(2.3 \times 10^4, 1.8 \times 10^{13})$, respectively, compared to the true mean $\bar{Y} = 1.7 \times 10^4$. Note the sensitivity of the interval to the choice of φ and the unfortunate fact that the interval for the better supported value of φ ($-1/4$) is very wide, yet fails to include $\bar{Y}$. As Rubin notes, the problem lies in the fact that the population mean is highly influenced by the right tail of the distribution. The model makes rather strong assumptions about this tail, which the observed data have very little power to address. If these assumptions are incorrect, then the model gives misleading answers. These problems are not shared by estimators of the population median, as Rubin demonstrates.

Estimation of the mean of a highly skewed population from a simple random sample is also considered by Hidiroglou and Srinath (1981), who consider mean-squared-error properties of estimators that downweight large units. They consider replacing the usual estimate of the total by

$$r \sum_{i=1}^{t} y_i + \frac{N - rt}{n - t} \sum_{i=t+1}^{n} y_i$$

where n is the sample size, N is the population size, the first t values of y_i are the outliers, and r is the weight given to the outliers. They consider $r = 1, r = T/t$, where T is the number of outliers in the population (a post-stratified estimator) and a value of r chosen to minimize mean squared error. The choice $r = 1$ assumes, in effect, that $T = t\,t$; that is, that all the outliers in the population appear in the sample, which seems unlikely given simple random sampling. The other estimators require some knowledge of T to be defined at all. Thus, all the estimators require or imply prior knowledge of the number of outliers in the population. Given such information, my preference would be to incorporate it into the analysis as a formal Bayesian prior, using the modeling framework adopted by Rubin (1983a), rather than restricting its use to ad hoc modifications of the usual estimator. Indeed, by incorporating mild prior information about bounds for the sizes of New York municipalities, Rubin (1983a) produced satisfactory Bayesian probability intervals for $\bar{Y}$ in his problem.

Since a simple random sample, in isolation, is ineffective for estimating means from very long-tailed populations, we should work hard to avoid being faced with this situation in practice. If prior estimates of size are available for the population, then unequal (for example, probability proportional to size) sampling can be used ensure that large units are included with high probability. If pps sampling is not possible, but a historical sequence of previous random samples is available, then the past can be used to build information about the incidence of outliers or other aspects of the distribution of interest. The incorporation of such prior information into the analysis of the current sample seems to be a topic where more research might be useful.

5.3 Accommodation: Robust Methods

In principle, I believe that statisticians should resist the temptation to tamper with data and should, as much as possible, analyze data as they arrive. Thus, while editing of blatant outliers that clearly do not belong in the population seems prudent, my inclination is to leave more marginal outliers in the data set without alteration. Standard methods based on normality may then be inappropriate, and robust procedures that limit the influence of outliers have a useful role, particularly now that computational constraints have been lessened by the availability of fast, cheap computing.

Clearly, I cannot do justice to the extensive literature on robust methods here; for reviews, see the recent texts by Huber (1981) and Hampel et al. (1986). The modeling approach to robust inference computes likelihood-based inferences for models involving long-tailed error distributions. For example, Box and Tiao (1973) consider Bayesian analysis of the linear model, replacing the usual normal error distribution with a distribution from the exponential power family. My own work has considered ML estimation assuming contaminated normal and t errors (Maronna, 1976; Rubin, 1983b; Pendergast and Broffitt, 1985; Sutradhar and Ali, 1986; Little, 1988c; Lange, Little, and Taylor, 1989).

These models have an interesting connection with the models for ratios in Section 4. Consider (5), (7), (8), and (10) with $\lambda = 1$, but suppose that the scale variable x_i is not observed but unobserved and is assumed to be *iid* over the observations with density $h(x)$. If νx_i is chi-squared with ν degrees of freedom, then, integrating over x_i, the marginal distribution of $\mathbf{z}_i$ is t distributed with unchanged mean, the scale is σ^2 in (5) and (7) or $\mathbf{\Sigma}$ in (8) and (10), and the degrees of freedom are ν. ML for these t models yields robust estimates of the parameters, with degree of downweighting controlled by the parameter ν. These estimates can be found by applying the EM algorithm and treating x_i as missing data, yielding an iterative version of the weighted-least-squares algorithms obtained when x_i is known (Dempster, Laird, and Rubin, 1977; Rubin, 1983b). An attraction of this approach is that ML estimation can be extended to handle missing data in the zs and that ν can itself be estimated given sufficient data; for details, see Little and Rubin (1987, Section 10.5), Little (1988c), and Lange, Little, and Taylor (1989).

Another choice of distribution for x_i gives a contaminated normal distribution for $\mathbf{Z}$, which may be useful when the outliers are not thought to be part of the population of interest. The EM algorithm for this model is again iteratively reweighted least squares, with a different form for the weights (Little and Rubin, 1987, Section 10.5). The contaminated normal model involves two additional parameters—the contamination fraction and the variance inflation factor—and, hence, involves more flexibility (and more decisions) than does the t model, which has just one additional parameter, ν.

EXAMPLE 2 In Example 1, the multivariate normal regression model (9) was fitted to resting hand-tremor data with $p = 8$ outcome variables $Y_1, \ldots, Y_8$ measuring activity in eight frequency bands and V a binary variable indicating TD and control groups. Since the plot in Figure 2 suggested longer than normal tails, we fitted a multivariate regression model with the normal replaced by the multivariate t with ν degrees of freedom. Maximized loglikelihood values for a grid of values of ν were as follows:

$df(\nu)$	∞	16	13	11	10	9	8	6
loglik, l	1513.4	1536.4	1537.8	1538.3	1538.4	1538.2	1537.5	1534.6

Thus, the model with $\nu = 10$ maximized the loglikelihood, with a significant improvement in fit over the normal model ($\chi_1^2 = 2l(\hat{\nu}) - 2l(\infty) = 50.0$, likelihood ratio test). Parameter estimates are not shown, but differ slightly from the normal model.

Residual distances from this model are again computed from (10) and (11), with $\hat{\mathbf{B}}$ and $\mathbf{S}$ replaced by the ML estimates of $\mathbf{B}$ and $\boldsymbol{\Sigma}$, d_i^2 restricted to the p_i observed variables in case i, and p in (11) replaced by p_i. For this t model, d_i^2/p_i and F_i have asymptotic F-distributions with p_i and ν degrees of freedom (Lange, Little, and Taylor, 1989). To make additional allowance for finite samples, I write

$$F_i \cong \frac{\chi_{p_i}^2/p_i}{(\chi_\nu^2/\nu)(\chi_{\nu_2}^2/\nu_2)} \cong \frac{\chi_{p_i}^2/p_i}{\chi_{\nu_3}^2/\nu_3}$$

where $\nu_2 = n - qp_i$, $\chi_{\nu_2}^2/\nu_2$ is the adjustment for finite samples introduced in the normal model, and $\nu_3 = \nu_2\nu/(\nu_2 + \nu + 2)$ is chosen so that $\chi_{\nu_3}^2/\nu_3$ has the same mean and variance as $(\chi_\nu^2/\nu)(\chi_{\nu_2}^2/\nu_2)$. Then, F_i is transformed to approximate normality using Eq. (8) with $\nu_1 = p_i$ and ν_2 replaced by ν_3.

The normal plot of these transformed residual distances is given in Figure 3 for the best-fitting t model with $\nu = 10$ degrees of freedom. Although not perfect, the data follow the line with unit slope more closely than in Figure 2, and the plot shows less evidence of outlying points.

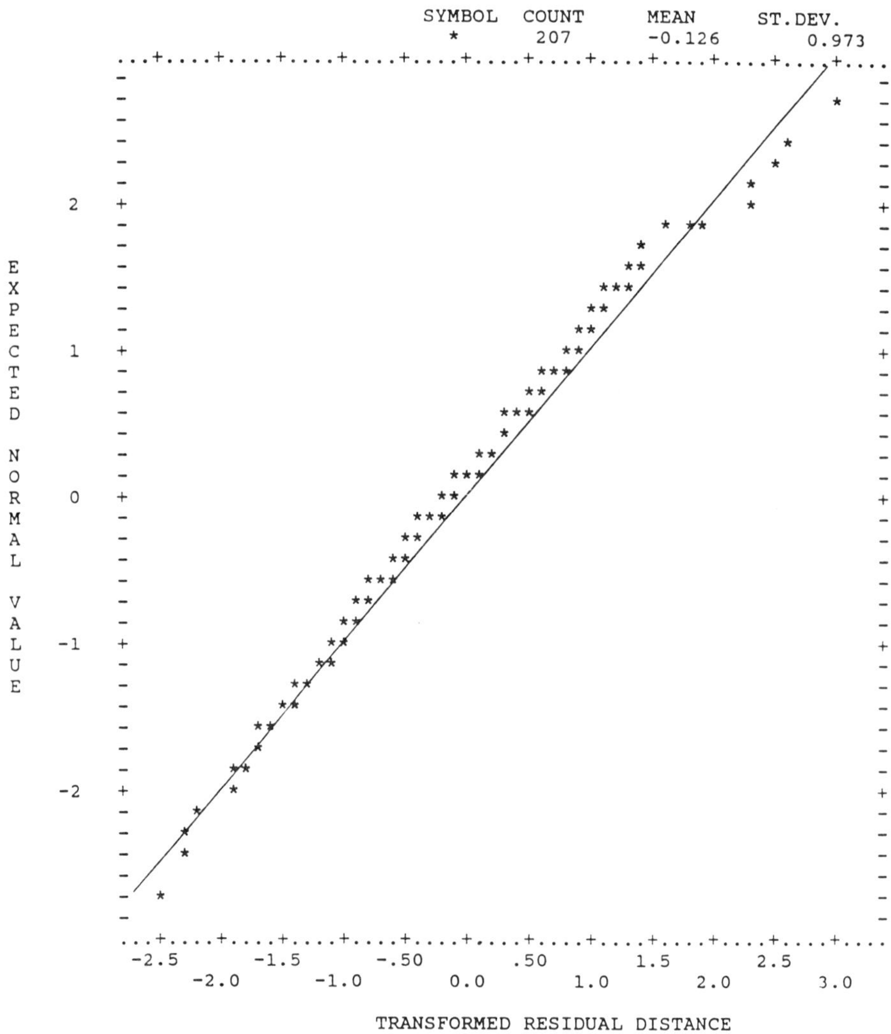

Figure 3. Normal plot of transformed residual distances, resting tremor data, T10 model.

6. IMPUTATION AND THE ALTERNATIVES

The standard approach to missing data created by the editing process is to replace edited values with imputed values—the terms edit and imputation seem to go hand in hand. However, imputation is not the only option, and its merits vary depending on the context. Of course, the attraction lies in the creation of a rectangular data set, amenable to standard statistical analyses. *Deductive* or *deterministic* imputation, where a missing value can be nailed down with negligible error from other data in the questionnaire, clearly seems appropriate. If missing values are less well determined, then a single impute cannot capture uncertainty in the imputation process, and its utility depends on the imputation method and the analytical context. Little (1988a) contains views on properties of a good imputation method.

A major defect of imputation is that analyses of the filled-in data have spurious precision, since error in the imputes is not properly reflected. This may not matter if point estimation is the major focus of the analysis, but it does if tests or confidence intervals are required. Rubin (1978) proposes *multiple imputation* as a solution to this problem. More than one (say, I) draw from the predictive distribution is imputed for each missing value, and the complete-data analysis is repeated I times, once with each imputation substituted. Let $\hat{\theta}_m$ be the estimate of a particular parameter θ from the mth analysis, and let $\hat{v}_m$ be the estimated variance. The final estimate of θ is $\hat{\theta} = \Sigma\hat{\theta}_m/I$, with estimated variance

$$\hat{v}^2 = s_w^2 + (1 + I^{-1})s_b^2$$

where $s_w^2 = \Sigma\hat{v}_m/I$ is the average variance within imputed data sets and $s_b^2 = \Sigma(\hat{\theta}_m - \hat{\theta})^2/(I - 1)$ is the between-imputation variance and reflects uncertainty in the imputation process. Large-sample inference for θ is based on treating $(\hat{\theta} - \theta)/\hat{v}$ as t distributed with $v = (I - 1)[1 + \{I/(I + 1)\}s_w^2/s_b^2]$ degrees of freedom. Theory underlying the method and examples are given in Rubin (1986, 1987).

An alternative to imputation is to leave data as missing, identified as such by a missing-data code. This approach defers the treatment of missing values to the data-analysis stage, where it rightly belongs. In principle, it allows the missing-data method to be tailored to the context of each specific analysis, as seems desirable. ML estimation with incomplete data is reviewed in Part II of Little and Rubin (1987); the approach is illustrated by the multivariate regression models with normal and t errors fitted as in Examples 1 and 2 above. Algorithms for ML estimation from incomplete data are becoming increasingly available; for example, the BMDP statistical package (Dixon 1987) now includes the multivariate normal EM algorithm in the BMDPAM program, and a powerful new program for repeated-measures analysis using data with missing values (BMDP5V), based on ML methodology in Jennrich and Schluchter (1985). Until such software becomes widely available, most data analysts will be restricted by available software to relatively naive missing-data methods such as complete-case or available-case analysis (Little and Rubin, 1987, Chapter 3), which may be worse than complete-data methods applied to carefully imputed data. Also, missing-data adjustments for a particular analysis will naturally tend to be based on the variables involved in that analysis and, hence, may neglect excellent predictors of the missing values that are not of immediate concern. Despite these reservations, leaving the data as missing has a certain integrity and is becoming an increasingly attractive option as efficient tools for handling incomplete data sets become more widely available.

Acknowledgments

This research was supported by grant MH 37188 from the U.S. National Institute of Mental Health. The author thanks Dr. William Wirshing for providing the data for Examples 1 and 2 and Janet Elashoff for making available preliminary analyses of these data.

REFERENCES

Anderson, T. W. (1958). *An Introduction to Multivariate Statistical Analysis*, New York: John Wiley & Sons.

Box, G. E. P. and Tiao, G. C. (1973). *Bayesian Inference in Statistical Analysis*, Reading, Mass.: Addison-Wesley.

Dempster, A. P., Laird, N. M., and Rubin, D. B. (1977). Maximum Likelihood from Incomplete Data via the EM Algorithm, *J of the Royal Statistical Society, B*, 39: 1–38.

Dixon, W. J., ed. (1987). *BMDP Statistical Software*, Berkeley: University of California Press.

Fellegi, I. P. (1975). Automatic Editing and Imputation of Quantitative Data (Summary), *Bulletin of the International Statistical Institute*, 46: 249–253.

Fellegi, I. P. and Holt, D. (1976). A Systematic Approach to Automatic Edit and Imputation, *J of the American Statistical Association*, 71: 17–35.

Gnanadesikan, R. (1977). *Methods for Statistical Data Analysis of Multivariate Observations*, New York: John Wiley & Sons.

Greenberg, B. (1982). Using an Edit System to Develop Editing Criteria, *Proceedings of the Survey Research Methods Section, American Statistical Association*, 1982: 366–371.

Hampel, F. R. et al. (1986). *Robust Statistics: An Approach Based on Influence Functions*, New York: John Wiley & Sons.

Hawkins, D. M. (1974). The Detection of Errors in Multivariate Data Using Principal Components, *J of the American Statistical Association*, 69: 340–344.

Hawkins, D. M. and Wixley, R. A. J. (1986). A Note on the Transformation of Chi-Squared Variables to Normality, *The American Statistician*, 40: 296–298.

Huber, P. J. (1981). *Robust Statistics*, New York: John Wiley & Sons.

Hidiroglou, M. A. and Berthelot, J.-M. (1986). Statistical Editing and Imputation for Periodic Business Surveys, *Survey Methodology*, 12: 73–84.

Hidiroglou, M. A. and Srinath, K. P. (1981). Some Estimators of Population Totals from Simple Random Samples Containing Large Units, *J of the American Statistical Association*, 76: 690–695.

Hinkins, S. and Scheuren, F. (1986). Hot Deck Imputation Procedure Applied to a Double Sampling Design, *Proceedings of the Survey Research Methods Section, American Statistical Association*, 1986: 181–195.

Jennrich, R. I. and Schluchter, M. D. (1986). Unbalanced Repeated-Measures Models with Structured Covariance Matrices, *Biometrics*, 42: 805–820.

Kendall, M. G. and Stuart, A. (1969). *The Advanced Theory of Statistics*, Vol. 1 (3rd ed.) New York: Hafner.

Lange, K., Little, R. J. A. and Taylor, J. M. G. (1989). Robust Statistical Inference Using the T Distribution, *J of the American Statistical Association* 84: 881–896.

Little, R. J. A. (1988a). Missing Data Adjustments in Large Surveys, *J of Business and Economic Statistics*, 6: 287–301 (with discussion).

Little, R. J. A. (1988b). Some Statistical Issues at the World Fertility Survey, *The American Statistician*, 42: 31–36.

Little, R. J. A. (1988c). Robust Estimation of the Mean and Covariance Matrix from Data with Missing Values, *Applied Statistics*, 37: 23–38. *Statistics.*

Little, R. J. A. and Rubin, D. B. (1987). *Statistical Analysis with Missing Data*, New York: John Wiley & Sons.

Little, R. J. A. and Schluchter, M. (1985). Maximum Likelihood Estimation for Mixed Continuous and Categorical Data with Missing Values, *Biometrika*, 72: 497–512.

Little, R. J. A. and Smith, P. J. (1987). Editing and Imputation for

Quantitative Survey Data, *J of the American Statistical Association*, 82: 58–68.

Madow, W. G., Nisselson, H., and Olkin, I., eds. (1983). *Incomplete Data in Sample Surveys, Volume 1: Report and Case Studies*, New York: Academic Press.

Maronna, R. A. (1976). Robust M-Estimators of Multivariate Location and Scatter, *Annals of Statistics*, 4: 51–67.

May, P. R. A., Lee, M. A., and Bacon, R. C. (1983). Quantitative Assessment of Neuroleptic-Induced Extrapyramidal Symptoms: Clinical and Nonclinical Approaches, *Clinical Neuropharmacology*, 6, *Suppl.* 1: S35–S51.

Pendergast, J. F. and Broffitt, J. D. (1985). Robust Estimation in Growth Curve Models, *Communications in Statistics Theory and Methods*, 14: 1919–1939.

Rubin, D. B. (1978). Multiple Imputation in Surveys—A Phenomenological Bayesian Approach to Nonresponse, *Proceedings of the Survey Research Methods Section, American Statistical Association*, 1978: 20–34.

Rubin, D. B. (1983a). A Case Study of the Robustness of Bayesian Methods of Inference: Estimating the Total in a Finite Population Using Transformations to Normality, in *Scientific Inference, Data Analysis and Robustness*, New York: Academic Press, pp. 213–244.

Rubin, D. B. (1983b). Iteratively Reweighted Least Squares, *Encyclopedia of the Statistical Sciences*, 4: 272–275.

Rubin, D. B. (1986). Basic Ideas of Multiple Imputation for Nonresponse, *Survey Methodology*, 12: 37–47.

Rubin, D. B. (1987). *Multiple Imputation for Nonresponse in Surveys*, New York: John Wiley & Sons.

Sande, G. (1979), Numerical Edit and Imputation, invited paper, 42*nd Session of the International Statistical Institute*, Manila, Philippines.

Sutradhar, B. C. and Ali, M. M. (1986). Estimation of Parameters of a Regression Model with a Multivariate T Error Variable, *Communications in Statistics Theory and Methods*, 15: 429–450.

Welniak, E. G. and Coder, J. F. (1980). A Measure of the Bias in the March CPS Earnings Imputation Scheme, *Proceedings of the Survey Research Methods Section, American Statistical Association*, 1980: 421–425.

11
Improving Data Quality in Network Surveys of Rare Populations

JOHNNY BLAIR* Survey Research Laboratory, University of Illinois at Urbana-Champaign, Urbana, Illinois

1. NETWORK SAMPLES

In recent years, there has been an increased use of network sample designs to address the problems of estimating the size of rare populations and locating those population members for interviewing (Sudman, Sirken, and Cowan, 1988). While network sampling is a method that can greatly improve the data quality of surveys of some rare populations, it can also introduce biases and errors of its own. This chapter addresses the proper application of network designs, focusing on the use of feasibility studies to determine whether a network sample design is appropriate, and, if so, to aid in the design decisions necessary to its proper use.

The quality of surveys of rare populations traditionally has been hampered in two areas, required sample sizes and underreporting by target population members. By definition, a rare target population is a small percentage of the general population, often 1 percent or less. In conventional designs, large samples are required to locate sufficient numbers of such populations even to estimate basic descriptive statistics; because of the small number of target population members on which they

**Current affiliation*: Survey Research Center, University of Maryland at College Park, College Park, Maryland.

are computed, these estimates are subject to high sampling variances. The magnitude of underreporting varies depending on the nature of the target population (and, of course, on the quality of the questionnaire and data-collection procedures). If the defining characteristic is sensitive and/or socially undesirable—e.g., drug use—the amount of underreporting may be substantial. Network sampling can be effective in addressing both these areas.

The basic idea of network sampling is quite simple and appealing; however, the situations in which it is both feasible and more efficient than a conventional household sample design are often not obvious. The primary difference between network surveys and conventional surveys is the number of enumeration units to which target respondents are linked. In conventional household sample surveys, each target respondent is linked to one enumeration unit: his or her place of residence. In network sampling, respondents are linked to (that is, can be reported by) additional enumeration units (Sirken, 1970). This linkage of target respondents to households other than their own is accomplished by using a multiplicity-counting rule. If the multiplicity-counting rule is an efficient one, the number of enumeration units (households) that can and will report members of the target population is greatly increased over the conventional method. At the same time, the probability nature of the sample design is retained, although all members of the target population do not have the same probability of selection.

In surveys designed to estimate the size of rare populations and/or to locate such population members for direct interview, network sampling may be a useful approach. Survey populations or population subdomains of interest may constitute 1 percent or less of the general population. Such target subgroups as households experiencing stranger abduction of a child, persons newly diagnosed for rare cancers or other health conditions, and very elderly persons with severe visual impairment have all been targets of recent surveys (Sudman, Sirken, and Cowan 1988). The differences between locating such target populations by conventional screening or through network sampling is illustrated in Figure 1. In a population of N households of which certain ones (designated T) have a characteristic of interest, the population is clearly divided between the proportion of households (π_T) with the characteristic and the proportion (π_0) without the characteristic. Any particular household, T, can be identified only if that household is contacted directly. The probability of locating a target household is simply the total number of such households ($N(\pi_T)$) over the total population N. If π_T is small, the sample size—and, hence, the cost—of locating target households is large and estimates of target population

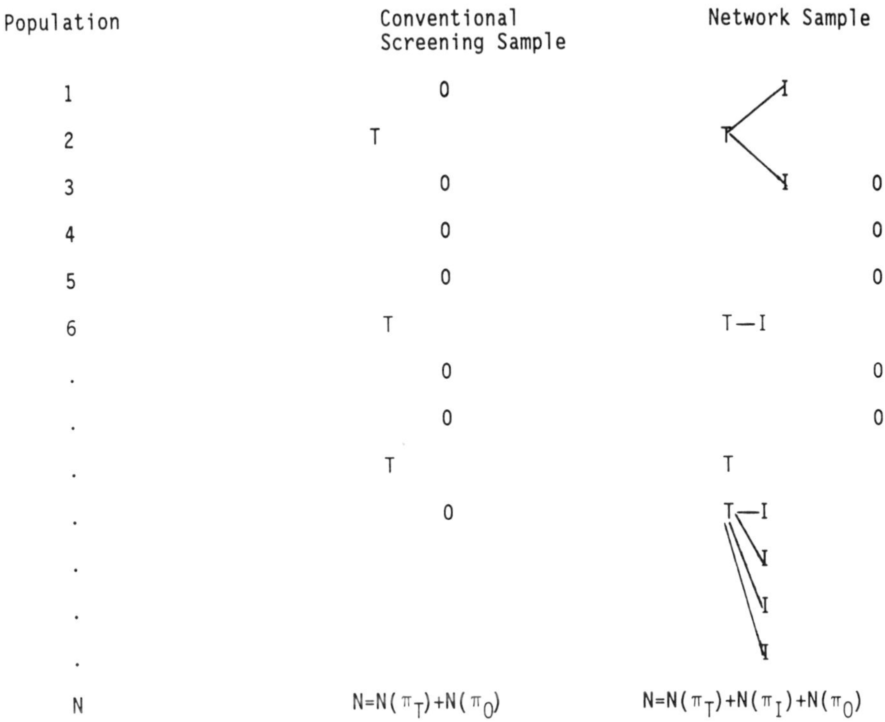

Figure 1. Comparison of Conventional and Network Sampling.

size and characteristics are subject to high sampling variances. In network sampling, a third group of households is defined as informant households and, while they may or may not be members of the target population themselves, are eligible to identify and report about target–population members. An informant household (designated in Figure 1 by *I*) may report a target household if it is linked to that household by a multiplicity-counting rule. The multiplicity-counting rule specifies the conditions that link the informant households to target-population households. The target-population households are eligible and required to be reported by the informant households to which they are linked by the rule, but are ineligible to be reported by any other units (Sirken, 1974). Examples of counting rules that have been used in several network feasibility and main studies are listed in Table 1.

The target household at the top of Figure 1 is linked to two informant households that may, in addition to itself, report that unit. This

Table 1. Examples of Multiplicity-Counting Rules

Three closest friends at work
Three closest nonwork friends
Parents in households other than respondent's
Siblings in households other than respondent's
Adult children in households other than respondent's
Neighbors on either side of the respondent's residence
Coworkers of the respondent

household is said to have a multiplicity of three. The second target household is linked to one informant household and, thus, has a multiplicity of two; the third target household is not linked to any informant household and, therefore, has a multiplicity of one, and so on. Several characteristics of network sample designs are immediately apparent from this simple illustration. First, the total number of households eligible to enumerate target population members increases from $N(\pi_T)$ to $N(\pi_T) + N(\pi_I)$. Second, the probabilities of any particular household being enumerated in the survey depends on the number of informant households to which it is linked by the multiplicity-counting rule. Therefore, the probabilities of selection of target households vary, producing a non-self-weighting sample. The network size (multiplicity) of each enumerated target household must be known to properly weight the sample results to produce unbiased estimates.

Although network samples greatly increase the efficiency of locating rare populations, there is some effect on sampling variances. Because households in a network must be weighted inversely by network size, the differential weights increase sampling variances. In many cases, however, the net effect is that the amount of information obtained reduces sampling variances far more than they are increased by the variability and weights (Sudman et al., 1987).

Clearly, there are many instances in which network sampling might be a reasonable approach for estimating the size of a rare target population, estimating population size and obtaining some other data about the key condition or event that defines that population, or obtaining information that allows the researchers to contact target-population members and interview them directly. However, the choice of appropriate counting rules, their comparative efficiency, and the ability of network members to accurately report the size of the networks must be determined before conducting such a project. When one adds to these factors the fact that questionnaire length—and, hence, cost—is increased by the ancillary data that must be collected as part of the network sample design, the value of feasibility pilot studies becomes clear (Rothbart, 1979).

2. THE ROLE OF FEASIBILITY STUDIES

For the method to be effective, the network members must know about the characteristics or events of interest and be willing to report them. Additionally, network members must be able either to report the network size of enumerated target households or provide contact information so that those households may be contacted directly to obtain the network size. Typically, multiplicity-counting rules have been based on various types of family relationships. Parents, siblings, and adult children in other households have all been used for this purpose. Other types of multiplicity-counting rules—such as coworkers, next-door neighbors, or close friends—have also been attempted but, in general, have proved to be more problematic. In some instances, the counting rule has also included the provision that network members must reside within a specified geographic area. For example, in a feasibility study to determine whether network sampling could be used to estimate the incidence of missing children, only network members living within the same metropolitan area as target households were defined as eligible to report (Sudman et al., 1987).

Before conducting a network sampling study, the extent to which designated network members outside the target individual's household might know about the characteristic of interest must be either inferred from other research or tested empirically in a feasibility study. In most instances, the latter course provides a sounder basis for deciding whether and how to proceed with the main study (Rothbart, 1979).

Often, several different counting rules may seem appropriate to use in a particular survey. Because, as noted above, additional questions related to the network aspect of the study add to its cost, the researcher's goal is to choose those counting rules that are most effective and to omit those for which either the yield or quality of reporting is poor. Also, as noted, the proportion of the population with the characteristic of interest, π_T, may or may not be known from other data.

The proportion of the population, π_I, eligible to nominate target households under a particular counting rule will usually not be known (unless an identical or very similar network study has been conducted) except from a feasibility study. Planning and budgeting a main study without an estimate of this latter proportion would be quite risky.

3. THE DESIGN OF FEASIBILITY STUDIES

There are two general approaches to the design of network-sampling feasibility studies: screening designs and seeded-sample designs. Which design or combination of designs is appropriate depends on whether the goal of the main study is simply to estimate the size of the target population or, as is more often the case, to obtain additional data about target-population members, either directly or from the network informants.

Each of these design approaches brings with it issues of data quality unique to network-sampling feasibility studies. These data-quality issues are crucial both to the proper conduct of the study and to the interpretation of the study results.

3.1 Screening Designs

In the screening design, a general population sample is selected and the households contacted to determine whether they are members of the defined target population. Additionally, each sample household is asked to enumerate those households linked to it by the counting rule, or rules, under investigation. The screening questions are repeated for each of the network members. Information necessary to contact any identified target households directly may, or may not, be asked as well. If interviews are attempted with these nominated households, the quality of the contact information and the accuracy of the informant report may both be assessed.

The screening design provides a direct comparison of conventional and network reporting for each of the specified counting rules. However, if the target population is sufficiently rare, very large sample sizes are required to generate enough informant reports to examine their accuracy.

In Figure 2, the basic steps involved in a screening design are shown. If, from other studies, one has some idea of the size of the target population, the sample size for the feasibility study may be based on this.

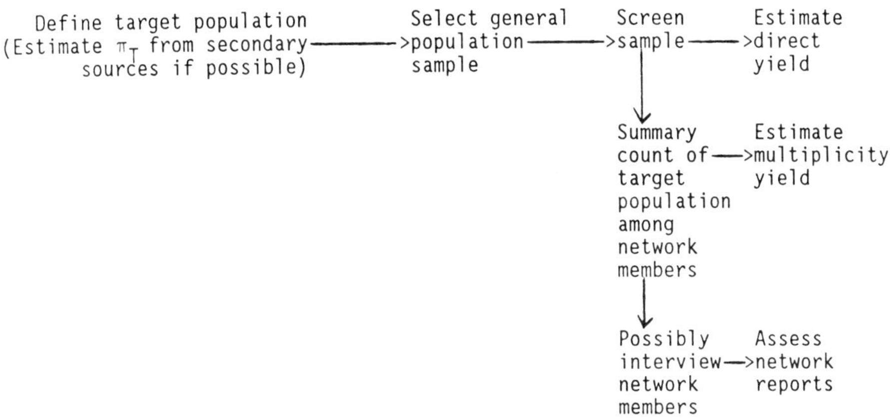

Figure 2. Screening-Sample Design.

Without this information, one runs the risk of enumerating too few of the target-population members to provide a good test of the method. If there is also interest in subdomains of the target population, the problem is increased.

Although simple screening designs are generally used less frequently than are seeded samples, an example of the appropriate use of such a design is worth noting. In preparing to conduct the Second Access to Health Care Study for the Robert Wood Johnson Foundation, network sampling was tested in a feasibility study* (Lakner and Sudman, 1985). The main project, a national telephone survey, required oversampling of adults and children who had been hospitalized in the past year or had one of a list of specified diseases or disabilities. The list included such rare conditions as cancer, epilepsy, and liver disease, as well as more common ones such as high blood pressure, asthma, and diabetes.

In a search of secondary sources on the prevalence of these special target subpopulations, it became clear that very substantial reductions in screening costs would occur if network sampling proved to be feasible.

The multiplicity-counting rules required asking the respondent about the parents, grandparents, grandchildren, and siblings of all house-

*Network Survey of Medical Costs. Principal Investigators Seymour Sudman (Survey Research Laboratory, University of Illinois) and Howard Freeman (UCLA). Funded by Robert Wood Johnson Foundation, 1985 (SRL Study No. 550).

Table 2. Yield of Target Population Members by Subgroup and Counting Rule

		Indirect (Multiplicity) Percent			
	Direct Percent	Children	Parents	Siblings	Grandparents
Hospitalization	26.0	6.7	29.2	28.4	8.1
High blood pressure	26.3	2.5	39.2	18.4	7.4
Asthma	10.4	2.4	7.8	6.5	1.6
Diabetes	7.7	0.8	12.0	7.1	2.4
Liver disease	0.9	—	1.1	0.6	0.2
Kidney disease	2.1	0.4	1.7	1.6	0.4
Epilepsy	1.1	0.3	1.2	1.5	—
Leukemia	0.2	—	0.5	—	0.2
Cancer	4.2	0.1	5.2	3.4	1.5
Not covered by insurance	11.5	3.6	5.6	11.4	—

hold members. A national RDD sample of approximately 1500 households was screened. The nominated target-population households were not contacted directly. The results are summarized in Table 2. The first column, *direct percent*, reports the percentage of sampled households that reported someone in their own household with the particular health condition. The *indirect percent* shows the percentage of sampled households that reported a network member, in another household, with the particular health condition. So, for example, 26 percent of the 1500 sample households reported that a household member was hospitalized overnight in the year preceding the interview; while 29.2 percent of the 1500 sample households reported that a parent, living in another household, was hospitalized in the preceding year. As anticipated, the use of network sampling adds substantially to the total expected yield.

The feasibility study also allowed sample estimates of yields by subgroup that might be expected—for alternative total sample sizes—in the main survey. Further, the efficacy of specific counting rules could be assessed. It was decided, for example, to limit screening about grandchildren to those grandchildren under 18 years old.

In the main study, the willingness of directly sampled households to give contact information was noticeably lower than in the feasibility study. This is most likely due to the much longer main study interview (approximately 15 vs. 45 minutes) and the unwillingness of respondents to expose family members (especially the very elderly) to a questionnaire of that length.

3.2 Seeded-Sample Designs

Seeded-sample designs provide an efficient way to measure directly both the amount and quality of reporting. In seeded-sample designs, one begins with a sample of known members of the target population. The source of this sample is generally a list or registry maintained by some agency. Obviously, such a list does not include all, or even most, members of the target population and is often very limited geographically, or the list would simply be used as the sample frame for the main study. A sample is selected from the list and added to or *seeded* into a general population sample. The general population sample, often designated as *decoys*, serves two purposes: First, to preserve the confidentiality of target households who may not report the defining event or characteristic; and second, to prevent the data-collection staff from expecting a target-household report in every interview. The former issue is especially important if the characteristic of interest is of a personal or sensitive nature. The second practice allows the researcher to simulate as closely as possible the conditions of a real network survey by preventing possible interviewer prompting for an expected report. In many cases, it has been found that a ratio of the seeded sample to the total sample of about 1 : 3 serves both purposes (Sudman et al., 1987). Since there will be some nonreporting of the target population among seeds and network members, the effective decoy sample size is increased.

In Figure 3, the steps involved in a typical seeded-sample design are presented schematically. As in the screening design, all interviewed households are asked for the names of network members required by the specified counting rules. For those households seeded into the sample, a sample of the network members listed for the household is selected and contacted for interview. The network members are administered a questionnaire identical to the one used for the seed and decoy samples. The seeded-sample design has the advantage that one knows at the outset of the study the amount of self-reporting to be expected from seeded households. Although one will typically compare network reports to self reports, this may not always be sufficient. For target populations in which the defining characteristic is sensitive, as well as rare (such as rape, certain types of cancer, or heroin use), one might expect that knowledgeable network members may report the event in some instances where it is not reported by the target-population member (Fishburne, 1980; Sirken, 1974). If this were the case, estimates of incidence based on combined self and

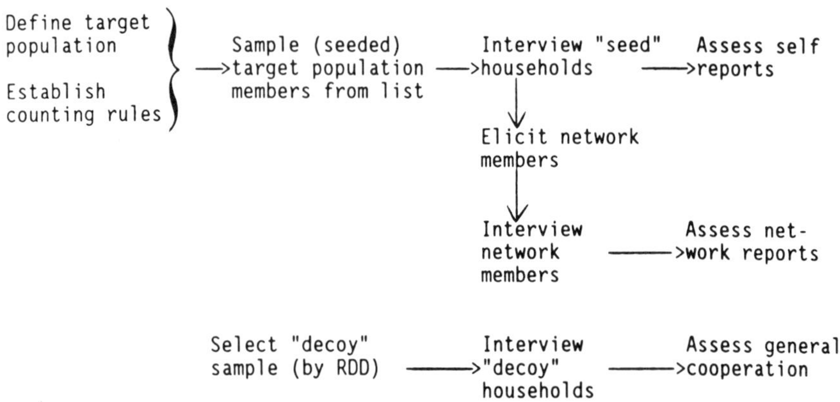

Figure 3. Seeded-sample design.

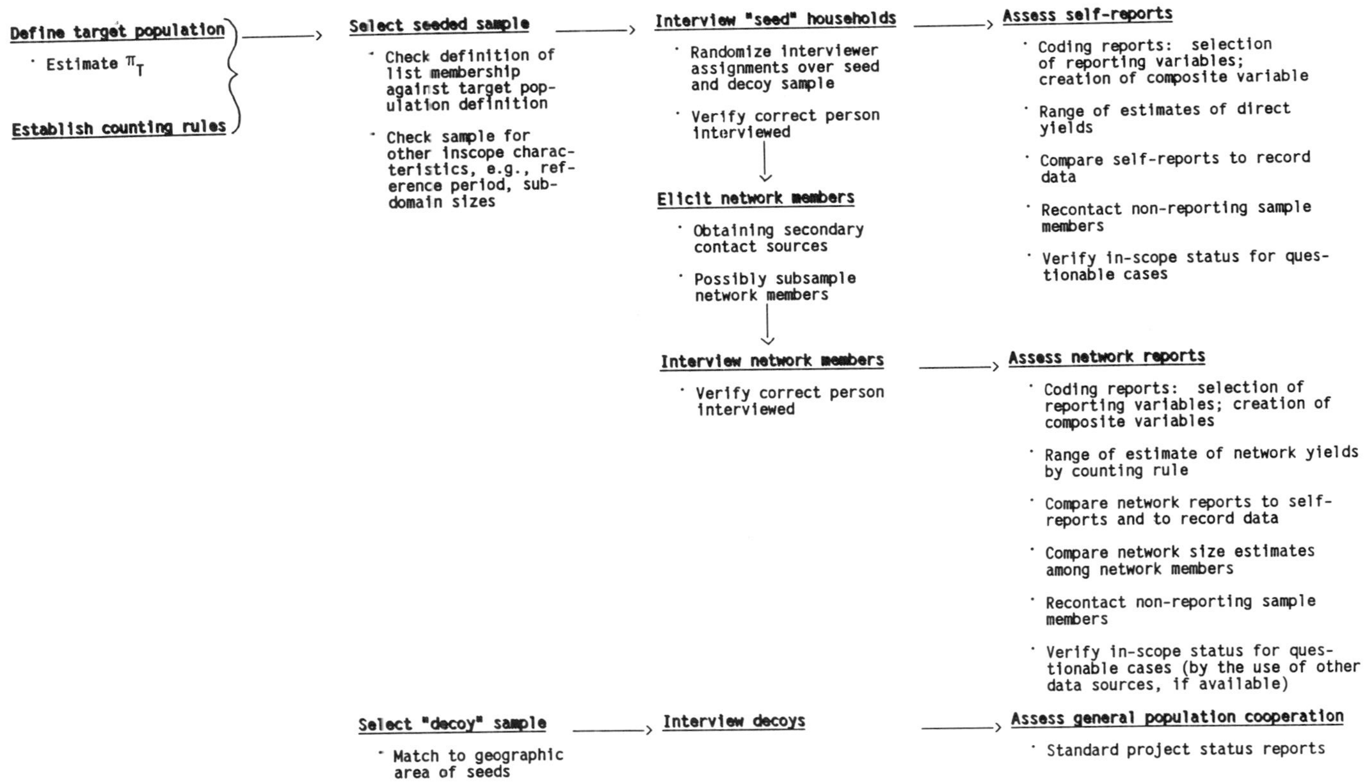

Figure 4. Data quality control points in seeded-sample designs.

network reports would be superior to estimates based on either type of report by itself, even if the amount of network reporting is low. A careful assessment of the presence of this phenomenon is possible using seeded-sample designs. Additionally, by requesting more nominations than will be used, the network sample that is generated can be subsampled to provide sufficient numbers of informant households to evaluate each counting rule. Further, comparisons of network sizes reported by target and informant households are also possible.

In Figure 4, the key data quality control points in the feasibility study are noted. These will be discussed in more detail in two case studies. Which of these items is of most concern to the researcher in terms of data quality depends on the purpose of the main study. In Table 3, the data required from informants are summarized by each of three general main study goals.

4. TWO SEEDED-SAMPLE STUDIES: INCIDENCE OF MISSING CHILDREN SURVEY AND LOCAL VICTIMIZATION SURVEY

To discuss the issues of data quality more concretely, two seeded-sample design case studies will be examined. One study employs a combined seeded-sample design and general population screening; the other is a simple seeded-sample design.

4.1 Incidence of Missing Children Survey

The purpose of this study* was to measure the accuracy with which specified relative networks will report missing children. The seeded sample was selected from state police I-SEARCH files. The study was conducted in the Chicago metropolitan area by telephone. The counting rules were: parents of the respondent, and (if applicable) of the respondent's spouse, living in separate households from the respondent, but in the Chicago metropolitan area; children, siblings, and grandparents of the respondent; coworkers; next-door neighbors. The sample sizes for the decoys, seeds, and generated network members are summarized in Table 4. After the elimination of nonresidential, nonworking, and other ineligible telephone

**Studying the Incidence of Missing Children by Special Survey Methods.* Principal Investigators, Seymour Sudman and Charles Cowan, Survey Research Laboratory, University of Illinois, Urbana-Champaign. Funded by U.S. Department of Justice, 1987 (SRL Study No. 602).

Table 3. Tasks Required of Informants

Main Study Objective	Know of Defining Characteristic(s)	Willing to Report	Know Additional Information about Target Person	Know Network Size	Provide Contact Information
Estimate size of target population	Yes	Yes	No	Yes	No
Estimate target population size and report key details of characteristics	Yes	Yes	Yes	Yes	No
Interview identified target population members	Yes	Yes	No	No	Yes

Table 4. Missing Children Survey: Sample Disposition

	Initial Sample	In-Scope Sample	Interviews	Eligible to Report	Percent Reporting Incident
Seeds	391	289	250	250	43.0
Generated network	160	133	125	125	10.3
RDD (decoys)	999	572	441	441	6.5*

*Including network and self reports.

numbers, an in-scope sample of 289 remained. At the time of initial contact, all of the in-scope sample of seeds (289) and generated network members (125) was considered eligible to report the missing child incident selected from the I-SEARCH records. As shown by the differences between *in-scope sample* and *eligible to report*, this, in fact, turned out not to be completely true.

A sample of 441 interviews was completed with decoy households selected using Waksberg-Mitofsky (Waksberg, 1978) random-digit dialing sampling. Since the same questionnaire was administered to all sample members, decoy households were asked to report about their own households and other households specified in the counting rules. The 6.5 percent of decoy household reports includes those reports for members of their networks. The network households nominated were not contacted directly.

4.2 Local Victimization Survey

The purpose of this feasibility study—conducted by telephone in the Peoria, Illinois, area—was to determine whether network sampling might be used to improve estimation and reporting of three types of criminal victimization: robbery, burglary, and assault (Blair and Czaja, 1987). These victimizations were chosen to include crimes against both households and individuals and to include at least one crime—assault—that is thought to be significantly underreported in traditional surveys. The seeded households were selected from records of the Peoria Police Department.

The counting rules examined were: three closest friends at work or three closest nonwork friends; and parents, siblings, and adult children living in separate households, all in the Peoria area. The sample sizes, interviews completed, and overall reporting percentages are listed in Table 5 for each of the three samples.

5. ISSUES OF DATA QUALITY

There are two general approaches that have been used to compare conventional and network sample designs or to compare alternative counting rules within the network approach. First, one may simply compare the yield of target-population households that may be achieved for a fixed cost. The second approach is to compare the mean square error for alternative counting rules and for conventional versus network

Table 5. Victimization Survey: Sample Disposition

	Sample	In-Scope	Interviews	Eligible to Report	Percent Reporting Incident
Seeds	1144	688	559	375	65.6
Generated network:					
Relatives	152	132	113	52	44.2
Friends	145	127	108	36	50.0
RDD (decoys)	161	124	82	—	—

samples (Nathan, 1976). Mean square error (MSE) is the sum of the variance and the square of the bias. In general, the bias component, when compared to the variance component of the mean square error, is smaller for both the conventional and multiplicity counting rules. However, the effect of the bias term relative to the variance increases as the sample size increases. Thus, a small bias becomes a dominant source of error as the sample size becomes large (Czaja et al., 1986; Kish, 1987).

The bias is the combination of reporting bias and sample bias. In most of the formulas developed to date (Czaja and Blair, 1988; Czaja et al., 1986), the bias term allows only for reporting bias. For the purposes of this descussion, reporting bias is understood to include nonreports of target households by informants eligible and required under the counting rule to report them (false negatives), as well households incorrectly identified as targets (false positives). Additionally, inaccurate reporting of the details of the target event or behavior contributes to bias. Clearly, these sources of bias can occur in both seed and network reports. By the very nature of the type of data required of network informants, reporting bias is a major data-quality concern. Before examining sources and possible controls to minimize the reporting bias, it is useful to note some issues of sample bias that may be addressed before data collection.

5.1 Sample Bias

The quality of the sample frame, important in any sample survey, may affect the results of a feasibility study in ways that are difficult to anticipate during planning. Seeded-sample designs, by their very nature, depend on the sampling frame being composed of true members of the target population. If frame members differ from the operational target-population definition, and these incorrect listings are not detected by the researcher, sample bias is introduced. Other incorrect information in the sampling frame—such as information needed to contact respondents—may, to the extent that it is not randomly distributed throughout the frame, add to the sample bias. In both these seeded-sample studies, substantial losses are noted between the initial sample size and the in-scope sample cases. Since, in many instances, one is asking about a characteristic or event that occurred during a specified time period, the number of seeded cases available is often quite limited. Losses due to the quality of the sampling frame (resulting from out-of-date listings or incorrectly classified households) may reduce the available in-scope sample below a usable level. Again, this problem may be especially severe for subdomains. In addition, in both of these studies, the listed phone number was sometimes that of a neighbor or relative, rather than of the target respondent. This situation, certainly not limited to the two lists involved here, makes it especially important to verify that the correct household has been reached (yet without alerting the field staff to the fact that a report is expected from a particular sample household). The incorrect inclusion of such households in the seeded sample would result in incorrectly classifying out-of-scope households as target households that did not report the event or condition.

The question of classifying households correctly as in-scope or not is also related to the basic definition of the target population. For example, the general term "missing children" includes several very different types of incidents, including runaways, parental abductions, acquaintance abductions, and stranger abductions. Since the effectiveness of network sampling might vary by the subcategories of the population definition, it is important to recognize these components in both the selection of the sample and the design of questionnaire items to elicit each type of incident. It was found that incidents classified by I-SEARCH as missing children events did not really fall under any of the definitions of interest in the project. Hence, the questionnaire was not designed to elicit reporting of some types of incidents, such as very short-term missing child occurrences. Again, the dual dangers of either underestimating the number of true reports or severely reducing the available seeded sample must be carefully avoided.

As part of the process of checking each element in the sampling frame, it must be recognized that the terms used for classification by the agency supplying the list are not necessarily defined the same as the researcher's understanding of those terms. In the victimization survey, for example, it was determined during the data collection and coding that a number of cases classified by the police department as assault were in fact verbal assaults. These incidents were threatened assaults that were reported to the police. The questionnaire was not designed to elicit this type of incident and, hence, reporting of it could not be expected. These cases were reclassified as out-of-scope. If data to supplement the sampling frame are available, they should be used to check carefully the reason for the inclusion and classification of sample members. In the victimization study, such data were available in separate narrative police reports on each crime incident. However, this was not determined until the coding phase of the project, rather than during preparation of the sampling frame (Czaja and Blair, 1988).

5.2 Reporting Bias

In assessing the accuracy of informant reports about the characteristic or event of interest, either the self reports of target-population households or record data from the seeded sample frame may be used as a basis for estimating reporting bias. As noted in Figure 4, either approach requires the selection of key items and the development of a classification scheme to determine whether the basic target-population definition has been met in the interview report and, further, how accurate are the details of the report (beyond the basic definitions). Since it is not realistic to expect perfect reporting by either informants or target respondents, some reasonable criteria of comparable reporting must be set. In the victimization survey, for example, key report items were listed (such as date of the victimization incident, whether a weapon was used, and amount of property loss or physical injuries), and some latitude for reporting error was allowed for each. The classification of a questionnaire as to whether it contained a correct report or not was transformed into a simple numerical composite variable based on these various items.

Another useful approach for assessing the amount of reporting that could be expected in a main study is to recontact households that were interviewed in the feasibility survey and did not report the target incident (whether seeded or informant households). In the missing children study, households that were seeded into the sample, yet did not report a missing child incident, were recontacted. In 94 cases, it was verified that the correct household had been reached. Of these 94 households, 65—upon direct questioning as to whether there was a missing child incident reported to the police within the reference period—admitted that such an incident had occurred, although they had not reported it in the original interview. Most commonly, the reason given for the initial nonreport was that the incident was considered too insignificant to report or the respondent simply did not want to go into the details of it. A finding like this, in the course of a feasibility study, has, of course, implications for questionnaire design and, perhaps, for interviewer training for the main study.

The willingness and ability of respondents to provide the information necessary to contact members of their network is, in many instances, crucial to the conduct of a network survey. In both of these studies, it was found that many respondents, while perfectly willing to complete an interview and report about themselves and members of their network, refused to give information necessary to directly contact network members. As noted above, this was also the experience in the Second Access to Health Care Survey. In some instances, respondents were willing to give the information for some network members but not others, while in other instances, all contact information was refused. The most common reason given was that the respondent did not feel he or she had the right to give out contact information for a network member, who would then be asked to participate in the survey. This tendency seems to vary considerably by type of counting rule, with respondents being much less willing to give out contact information for elderly relatives or for neighbors than for younger relatives or close friends. Although this problem may be somewhat alleviated by the vigorous use of any available secondary sources, this differential cooperation by counting rule should certainly be a factor in assessing the feasibility of using a particular rule in a main survey.

The last item of interest related to reporting bias is the ability of both seeded and informant respondents to place an event within a specified reference period. The traditional concerns of omissions and telescoping are also factors in the quality of data from network surveys. That is, respondents may not report events or behaviors that occurred within the reference period, or may incorrectly report events and behaviors that actually occurred before the reference period. Again, care must be taken that similar latitude for error is allowed for both informant and seeded-respondent reporting, so that the expectations of accuracy in placing events within reference periods are based on the same standard.

6. RECOMMENDATIONS

In developing an approach for improving data quality in network-sampling feasibility studies, one notes certain patterns of data-quality problems over the studies discussed in this chapter, as well as in several other studies. While the various contributions of different components of sample and reporting bias to the total mean square error are not readily quantifiable during the planning of the feasibility study, the sources of such bias do appear to be somewhat limited. A number of steps can be taken to minimize the effect of these sources of bias during the preparation of the sampling frame and questionnaire design. If, in addition, special care is taken to recognize the special nature of such feasibility studies in the coding and classification of the survey data, significant improvements in overall data quality are obtainable.

Several suggestions for the conduct of these types of studies are worth consideration as future research procedures.

1. Recognize and stratify seeded samples by subpopulation. Oversample subpopulations of interest to ensure sufficient numbers of completed cases to assess the results for subdomains.
2. Use all available information from records of seeded-sample members to assess whether the unit is part of the defined target population and should be expected to be reported in the questionnaire.
3. Consider the use of focus groups or small pretests to limit the number of counting rules to be tested in the feasibility study, so that the resulting questionnaire is reasonably comparable in length and burden to that which would be used in the main study.
4. Consider recontacting households where a report was expected but not forthcoming, to determine the reason for the nonreport. (Special care must be taken here to preserve the confidentiality of respondents and not to affect the data-collection operation.)
5. In attempting to determine the quality of contact information provided by respondents, use the information provided in the interview to attempt to contact the named respondents, rather than simply comparing the interview information to that contained in the seeded-sample record.
6. Consider the use of "snowball" sampling methods to generate additional secondary sources if necessary to track identified target respondents. This may prove especially effective when the respondent is willing to give contact information for some network members, but not for those randomly selected for interview. Perhaps the members made available can supply contact information for those randomly designated for interviewing.
7. Take into account during questionnaire design and coding-scheme development whether the expected direction of reporting bias would be toward false positives or false negatives.

7. AREAS FOR FURTHER RESEARCH

Further research in the development of simplified models for comparing alternative counting rules would be useful. The comparison of mean square errors, while statistically sound, requires the development of complex estimators of both sampling variances and reporting errors, as well as relatively sophisticated programming to quantify the feasibility study results. The development of simplified, yet defensible, models for comparing alternative counting rules would be very useful for many small-scale research efforts.

Network sample design research would also benefit from the development of simulation models that estimate π_I for particular counting rules. Even though the model would not address issues of the amount or quality of reporting, it might be used (in conjunction with reporting experiences for similar variables) to determine whether conducting a feasibility study is justified and how such a study might be designed.

REFERENCES

Blair, J. and Czaja, R. (1987). *Using Network Sampling for Local Victimization Surveys: An Evaluation of Reporting*, Midwest Sociological Society, Chicago.

Czaja, R. and Blair, J. (1988). *Evaluating Network Sampling for Local Victimization Surveys: Final Report*, Grant No. 85-IJ-CX-0032, National Institute of Justice.

Czaja, R. and Blair, J. (1988). Using Network Sampling for Rare Populations: An Application to Local Crime Victimization Surveys. *Proceedings of the Survey Research Section, American Statistical Association.*

Czaja, R., Snowden, C. B., and Casady, R. J. (1986). Reporting Bias and Sampling Error in a Survey of a Rare Population Using Multiplicity Counting Rules. *J of the American Statistical Association*: 411–419.

Fishbourne, P. (1980). *Survey Techniques for Studying Threatening Topics: A Case Study on the Use of Heroin*, Dissertation, Dept. of Sociology, New York University.

Kish, L. (1987). *Statistical Design for Research.* John Wiley & Sons, New York.

Lakner, E. and Sudman, S. (1985). *Pilot Study for Second National Access to Health Care Survey*, Project No. 550 Internal Report, Survey Research Laboratory, University of Illinois at Urbana-Champaign.

Nathan, G. (1976). An Empirical Study of Response and Sampling Errors for Multiplicity Estimates with Different Counting Rules. *J of the American Statistical Association in 1976*: 808–815.

Rothbart, George S. (1979). Discussion: Network Sampling in Health Surveys. *Proceedings of the Third Biennial Conference on Health Survey Research Methods*, Reston, Virginia, May 1979.

Sirken, Monroe G. (1970). Household Surveys with Multiplicity. *J of the American Statistical Association, 65*: 257–66.

Sirken, Monroe G. (1974). The Counting Rule Strategy in Sample Surveys. *Proceedings of the Survey Research Section, American Statistical Association*, pp. 119–123.

Sudman, S., Cowan, C., Blair, J., and Khodadadi, K. (1987). *Studying the Incidence of Missing Children by Special Survey Methods: Final Report*, Grant No. 86-ML-CX-0002, U.S. Department of Justice.

Sudman, S., Sirken M. G., and Cowan, C. D. (1988). Sampling Rare and Elusive Populations. *Science, 240*: 991–996.

Waksberg, Joseph, (1978). Sampling Methods for Random Digit Dialing. J of the American Statistical Association, *73, no.* 361: 40–46

12

Regression Estimators with Subsampling of Nonrespondents

PODURI S. R. S. RAO University of Rochester, Rochester, New York

1. INTRODUCTION

Consider a finite population of size N and a random sample of size n drawn without replacement from it. Frequently, in surveys on human populations, only n_1 of the sampled units respond on an item of interest y, but all of them may respond on a supplementary variable x. In household surveys, income-related items, expenditures, and consumptions are of the first type; family size and years of education are of the second type. In industrial and commercial surveys, production, quality of the output, etc., are of the first type; employee size, allocation of the amount of time for specific purposes, etc., are of the second type.

When nonresponse occurs, the estimators for the population mean and total based on the responses alone are biased. Hansen and Hurwitz (1946) suggest the possibility of obtaining information on the item of interest y on a subsample of the nonresponding units through intensive efforts. An unbiased estimator for the population mean of y is obtained by properly weighting the sample means obtained at the two stages. This estimator, its variance, and the sample size required to minimize the variance of the estimator or the total cost for the survey are presented in Section 2.

When there is complete response on the item of interest y and the supplementary variable x, the regression estimator for the population mean of y is known to have smaller variance than the sample mean. In this chapter, we combine the above subsampling procedure and the regression methodology to obtain estimators with small variance. Two estimators for the regression coefficient are suggested in Section 3.

For the procedure of subsampling the nonrespondents, two regression estimators for the population mean are examined in Section 4. The second estimator uses the information on x available from all the n units. The results indicate that it may not, however, have smaller variance than the first estimator, which uses only the information on x obtained from the initial responses and the subsampled units.

In Section 5, the population mean of y is expressed as the weighted average of the means of the response and nonresponse strata, and two more estimators for the population mean are obtained by combining the regression methodology and the subsampling procedure. The second estimator proposed in this section is seen to have smaller variance than the three remaining estimators.

Section 6 contains the procedures for determining the optimum sample sizes for each of the four estimators. Illustrations of the performances of the four estimators and the optimum sample sizes are presented in Section 7. Important conclusions of the entire study are presented in Section 8.

The methodology for the above estimators is somewhat similar to the one considered by the author in Rao (1986) for the ratio estimation. Special attention, however, is given to the estimation of the regression coefficient. The regression estimators in this chapter are derived by adjusting for the biases in the estimators suggested by the author in Rao (1987).

DEFINITIONS The population mean and variance of y are

$$\bar{Y} = \sum_{1}^{N} y_i/N$$

and

$$S_y^2 = \sum_{1}^{N} (y_i - \bar{Y})^2/(N-1)$$

Similarly, the mean and variance of x are

$$\bar{X} = \sum_1^N x_i/N$$

and

$$S_x^2 = \sum_1^N (x_i - \bar{X})^2/(N-1)$$

The covariance between x and y is

$$S_{xy} = \sum_1^N (x_i - \bar{X})(y_i - \bar{Y})/(N-1)$$

and the correlation coefficient is $\rho = S_{xy}/S_x S_y$.

In the case of nonresponse, the population is supposed to be consisting of N_1 respondents and N_2 nonrespondents. Let $(\bar{X}_1, \bar{Y}_1)$, (S_{x1}^2, S_{y1}^2), and S_{xy1} denote the means, variances, and covariance of the N_1 respondents. Similarly, let $(\bar{X}_2, \bar{Y}_2)$, (S_{x2}^2, S_{y2}^2), and S_{xy2} denote the corresponding quantities for the N_2 nonrespondents. With this notation, the population means can be written as $\bar{Y} = W_1\bar{Y}_1 + W_2\bar{Y}_2$ and $\bar{X} = W_1\bar{X} + W_2\bar{X}_2$, where $W_1 = (N_1/N)$ and $W_2 = (N_2/N)$. For the estimation procedures in this chapter, the N_1 respondents are considered to be those persons who will respond if they were drawn into the initial sample. The second group of N_2 persons are the ones who will not respond at the first attempt, but will respond if they were selected again in the second sample.

The means, variances, and covariance of the n sampled units are

$$\bar{y} = \sum_1^n y_i/n$$

$$\bar{x} = \sum_1^n x_i/n, \quad s_y^2 = \sum_1^n (y_i - \bar{y})^2/(n-1)$$

$$s_x^2 = \sum_1^n (x_i - \bar{x})^2/(n-1)$$

and

$$s_{xy} = \sum_1^n (x_i - \bar{x})(y_i - \bar{y})/(n-1).$$

When nonresponse occurs, let $(\bar{x}_1, \bar{y}_1)$, $(s_{x_1}^2, s_{y_1}^2)$, and s_{xy1} denote the means, variances, and covariance of the n_1 respondents. Similarly, let $(\bar{x}_2, \bar{y}_2)$, $(s_{x_2}^2, s_{y_2}^2)$, and s_{xy2} denote the corresponding quantities of the $n_2 = (n - n_1)$ nonrespondents.

2. HANSEN AND HURWITZ'S ESTIMATOR

Hansen and Hurwitz (1946) recommend drawing a subsample of size $m = n_2/k$, $k \geqslant 1$, from the n_2 nonrespondents and obtaining the mean $\bar{y}_m = \Sigma_1^m y_i/m$. The estimator for $\bar{Y}$ suggested by these authors is

$$\bar{y}^* = w_1\bar{y}_1 + w_2\bar{y}_m \tag{1}$$

where $w_1 = (n_1/n)$ and $w_2 = (n_2/n)$.

With I denoting the initial sample and the responses, $E(\bar{y}_m I) = \bar{y}_2$.
where $s_{e2}^2 =$

$$V(\bar{y}^*) = \frac{(1-f)}{n} S_y^2 + W_2 \frac{(k-1)}{n} S_{y2}^2 \tag{2}$$

where $f = (n/N)$ is the sampling fraction.

2.1 Optimum Sample Sizes

The cost of sampling is usually considered to be

$$C = c_0 n + c_1 n_1 + c_1 m \tag{3}$$

In this expression, C is the total available amount, c_0 is the cost per unit of sampling at the initial attempt, c_1 is the cost per unit of processing the returns of the respondents of the initial attempt, and c_2 is the cost per unit of obtaining responses from the subsampled units and processing them. The expected cost is

$$C = \left(c_0 + c_1 W_1 + \frac{c_2 W_2}{k}\right) n \tag{4}$$

As shown by Cochran (1977, p. 372), the value of k for minimizing the variance in (2) for a given expected cost or for minimizing the expected cost for a given variance is obtained from

$$k_{\text{opt}}^2 = (t - W_2)q \tag{5}$$

where $t = (S_y^2/S_{y_2}^2)$ and $q = c_2/(c_0 + c_1 W_1)$. With this value of k, from (2), the optimum value of n for a given variance V is

$$n_{\text{opt}} = \frac{N}{NV + S^2}[S^2 + W_2(k_{\text{opt}} - 1)S_2^2] \tag{6}$$

For a given C, the optimum sample size is

$$n_{\text{opt}} = \frac{C}{c_0 + c_1 W_1 + c_2 W_2/k_{\text{opt}}} \tag{7}$$

Note that if there is no nonresponse, the size of the initial sample for a given cost is $n = C/(c_0 + c_1)$. Since n_{opt} in (6) should not exceed this value, $(c_1 W_1 + c_2 W_2/k) > c_1$; that is, c_2 should not exceed kc_1.

An alternative procedure for finding the sample sizes was suggested by Srinath (1971) and J. N. K. Rao (1973), and a discussion of this procedure was given in Rao (1983).

2.2 Estimation of the Variance

Let

$$s_{ym}^2 = \sum_1^m (y_i - \bar{y}_m)^2/(m - 1)$$

denote the variance of the m sybsampled units. Since

$$E(s_{ym}^2|I) = s_{y2}^2$$

this variance is unbiased for S_{y2}^2. The sample variance s_y^2 is unbiased for S_y^2, but it is not available due to the nonresponse of the n_2 units. To obtain an unbiased estimator for S_y^2, note that

$$\begin{aligned}(n - 1)s_y^2 &= \sum_1^n y_i^2 - n\bar{y}^2 \\ &= \sum_1^n y_i^2 - n(w_1^2\bar{y}_1^2 + w_2^2\bar{y}_2^2 + 2w_1 w_2 \bar{y}_1 \bar{y}_2)\end{aligned} \tag{8}$$

An unbiased estimator of $\Sigma_1^n y_i^2$ is $\Sigma_1^{n_1} y_i^2 + (n_2/m)\Sigma_1^m y_i^2$. Further,

$$\begin{aligned}E(\bar{y}_m^2|I) &= \bar{y}_2^2 + V(\bar{y}_{2m}|I) \\ &= \bar{y}_2^2 + \frac{n_2 - m}{n_2 m} s_{y2}^2\end{aligned} \tag{9}$$

Since s_{ym}^2 is unbiased for s_{y2}^2, from (8), an unbiased estimator for $\bar{y}_2^2$ is $\bar{y}_m^2 - (n_2 - m)s_{y2}^2/n_2 m$. Substituting these results in (8), an unbiased estimator for S_y^2 is obtained from

$$\begin{aligned}(n-1)\hat{S}_y^2 &= \sum_1^{n_1} y_i^2 + (n_2/m)\sum_1^m y_i^2 - n[w_1^2\bar{y}_1^2 + w_2^2\bar{y}_m^2 \\ &\quad - w_2^2(n_2 - m)s_{ym}^2/n_2 m + 2w_1 w_2 \bar{y}_1 \bar{y}_m] \\ &= \sum_1^{n_1} y_i^2 + (n_2/m)\sum_1^m y_i^2 - n\bar{y}^{*2} + w_2(k-1)s_{ym}^2/n \\ &= (n_1 - 1)s_{y1}^2 + (n_2/m)(m-1)s_{ym}^2 \\ &\quad + nw_1 w_2(\bar{y}_1 - \bar{y}_m)^2 + w_2(k-1)s_{ym}^2/n \end{aligned} \tag{10}$$

Since $E(\hat{S}_y^2|I) = s_y^2$, the variance in (10) is unbiased for S_y^2. The estimator of $V(\bar{y}^*)$ is obtained by substituting $\hat{S}_y^2$ for S_y^2 and s_{ym}^2 for S_{y2}^2. Earlier, expressions for the estimators of $V(\bar{y}^*)$ through different approaches were obtained by Cochran (1977, p. 333), J. N. K. Rao (1973), and Rao (1986).

3. ESTIMATION OF THE REGRESSION COEFFICIENT

If there is complete response, an estimator of the population regression coefficient $\beta = S_{xy}/S_x^2$ is $b = s_{xy}/s_x^2$. With subsampling the nonrespondents, unbiased estimators for S_x^2 and S_{xy}, similar to $\hat{S}_y^2$ in (9), are obtained from

$$\begin{aligned}(n-1)\hat{S}_x^2 &= (n_1 - 1)s_{x1}^2 + (n_2/m)(m-1)s_{xm}^2 \\ &\quad + nw_1 w_2(\bar{x}_1 - \bar{x}_m)^2 + w_2(k-1)s_{xm}^2/n \end{aligned} \tag{11}$$

and

$$\begin{aligned}(n - I)\hat{S}_{xy} &= (n_1 - 1)s_{xy1} + (n_2/m)(m-1)s_{xym} \\ &\quad + nw_1 w_2(\bar{x}_1 - \bar{x}_m)(\bar{y}_1 - \bar{y}_m) + w_2(k-1)s_{xym}/n \end{aligned} \tag{12}$$

where $s_{xm}^2 = \Sigma_1^m (x_i - \bar{x}_m)^2/(m-1)$ and $s_{xym} = \Sigma_1^m (x_i - \bar{x}_m)(y_i - \bar{y}_m)/(m-1)$. An estimator for β is

$$b^* = \frac{\hat{S}_{xy}}{\hat{S}_x^2} \tag{13}$$

Since $E(b^*|I) \doteq b$, this estimator will be unbiased for β for large samples.

If w_2 is small, $\hat{S}_x^2 \doteq (n_1 - 1)s_{x1}^2 + (n_2 - 1)s_{xm}^2$ and $\hat{S}_{xy} \doteq (n_1 - 1) \times s_{xy1} + (n_2 - 1)s_{xym}$. These approximations were considered in Rao (1987) for estimating β.

If there is nonresponse on y but complete response on x, an alternative estimator for β is

$$b^{**} = \frac{\hat{S}_{xy}}{s_x^2} \tag{14}$$

Since $E(b^*|I) \doteq b^{**}$, for large samples, the variance of b^* will be larger than the variance of b^{**}. Both of them are, of course, biased for β for small samples.

The regression estimators for $\bar{Y}$ in the following section are based on b^*. There may be some advantage in replacing b^* by b^{**} in the expressions of these estimators. A similar procedure in the context of the ratio estimation was examined in Rao (1987). The results obtained there suggest that replacing b^* by b^{**} will reduce the bias of the regression estimator of $\bar{Y}$ for small samples, but it may not reduce the mean square error of the estimator unless the correlation between x and y is small.

4. REGRESSION ESTIMATORS FOR THE MEAN

In this section, we suggest two estimators for $\bar{Y}$ and compare their variances. The second estimator uses, to some extent, the observations on x of all the n units of the first sample.

When only n_1 of the n units respond on both x and y and when m units are subsampled from the n_2 nonrespondents, a regression type of estimator for $\bar{Y}$ is

$$l_1 = \bar{y}^* + b^*(\bar{X} - \bar{x}^*) \tag{15}$$

Since

$$E(l_1|I) \doteq \bar{y} + b(\bar{X} - \bar{x}) \tag{16}$$

$$V[E(l_1|I)] \doteq \frac{(1-f)}{n} S_e^2 \tag{17}$$

where $S_e^2 = S_y^2(1 - \rho^2) = \Sigma_1^N[(y_i - \bar{Y}) - \beta(x_i - \bar{X})]^2/(N - 1) = S_y^2 + \beta^2 S_x^2 - 2\beta S_{xy}$. Further,

$$V(l_1|I) \doteq w_2^2 \frac{(n_2 - m}{n_2 m} s_{e2}^2 \tag{18}$$

and

$$E[V(l_1|I)] = \frac{W_2(k-1)}{n} S_{e2}^2$$

where $s_{e2}^2 = \Sigma_1^{n_2} [(y_i - \bar{y}_2 - b^* (x_i - \bar{x}_2)]^2/(n_2 - 1) = s_{y2}^2 + b^{*2} s_{x2}^2 - 2b^* s_{xy2}$ and $S_{e2}^2 = \Sigma_1^{N_2}[(y_i - \bar{Y}_2) - \beta (x_i - \bar{x}_2)^2]/(N_2 - 1) = S_{y2}^2 + \beta^2 S_{x2}^2 - 2\beta S_{xy2}$.

From (17) and (18), an approximation to the variance of l_1 is

$$V_1 \doteq \frac{(1-f)}{n} S_e^2 + \frac{W_2(k-1)}{n} S_{e2}^2 \tag{19}$$

An approximate estimator for V_1 is obtained by replacing S_e^2 by

$$\hat{S}_e^2 = \frac{n-1}{n-2} (\hat{S}_y^2 - b^{*2} \hat{S}_x^2)$$

and S_{e2}^2 by

$$s_{em}^{*2} = \frac{\Sigma_1^m [(y_i - \bar{y}_m) - b^*(x_i - \bar{x}_m)]^2}{m-2}$$

Complete Response on *x* But Not on *y*

In this case, an estimator for $\bar{Y}$ is

$$l_2 = \bar{y}^* + b^*(\bar{X} - \bar{x}) \tag{20}$$

As in the case of l_1, $E(l_2|I) \doteq \bar{y} + b(\bar{X} - \bar{x})$. The large sample variance of l_2 is

$$V_2 \doteq \frac{(1-f)}{n} S_e^2 + \frac{W_2(k-1)}{n} S_{y2}^2 \tag{21}$$

An approximate estimator of V_2 is obtained by replacing S_e^2 by $\hat{S}_e^2$ above and S_{y2}^2 by s_{ym}^2.

As mentioned in the previous section, b^* in (15) and (20) may be replaced by b^{**}. But the gains due to such a procedure may not be significant.

From (19) and (21),

$$V_1 - V_2 = \frac{W_2(k-1)}{n} \beta^2 S_{x2}^2 \left[1 - 2\frac{\rho_2}{\rho}\frac{(S_{y2}/S_y)}{(S_{x2}/S_x)}\right] \tag{22}$$

where $\rho_2 = S_{xy2}/S_{x2}S_{y2}$. Although the variances of x and y of the entire population will be larger than the variances of the second stratum, the difference in the ratios (S_{y2}/S_y) and (S_{x2}/S_x) may not be significant. In that case, V_2 will be smaller than V_1, provided ρ_2 is smaller than ρ. We note, however, that if ρ_2 is small, the value of V_2 will be large and the regression procedure will not be beneficial.

5. TWO MORE ESTIMATORS

These estimators are obtained by writing $\bar{Y}$ as $W_1\bar{Y}_1 + W_2\bar{Y}_2$ and finding estimators for $\bar{Y}_1$ and $\bar{Y}_2$. One possible estimator for $\bar{Y}_2$ is $\hat{\bar{Y}}_2 = \bar{y}_m + b_{2m}(\bar{x}_2 - \bar{x}_m)$, where $b_{2m} = (s_{xym}/s_{xm}^2)$. Hence, an estimator for $\bar{Y}$ is

$$l_3 = w_1\bar{y}_1 + w_2[\bar{y}_m + b_{2m}(\bar{x}_2 - \bar{x}_m)] \tag{23}$$

Since $E(\hat{\bar{Y}}_2|I) \doteq \bar{y}_2$, $E(l_3|I) \doteq \bar{y}$ and, hence, $V[E(l_3|I)] = (1-f)S_y^2/n$. Further,

$$\begin{aligned} V(l_3|I) &\doteq w_2^2\frac{(n_2-m)}{n_2m}\frac{\Sigma_1^{n_2}[(y_i-\bar{y}_2)-b_2(x_i-\bar{x})]^2}{n_2-1} \\ &= \frac{w_2(k-1)}{n}s_{y2}^2(1-r_2^2) \end{aligned} \tag{24}$$

where $b_2 = (s_{xy2}/s_{x2}^2)$ and $r_2 = (s_{xy2}/s_{x2}s_{y2})$. From these results, an approximation to the variance of l_3 is

$$V_3 \doteq \frac{(1-f)}{n}S_y^2 + \frac{W_2(k-1)}{n}S_{y2}^2(1-\rho_2^2) \tag{25}$$

Comparison of V_1 in (19) with V_3 in (25) shows that l_1 will have smaller variance than l_3, unless ρ is considerably smaller than ρ_2.

An estimator of V_3 is obtained by replacing S_y^2 in (25) by $\hat{S}_y^2$ and $S_{y2}^2(1-\rho_2^2)$ by

$$s_{em}^2 = \frac{\Sigma_1^m[(y_i-\bar{y}_m)-b_{2m}(x_i-\bar{x})]^2}{m-2}$$

A Modified Estimator

The above estimator does not make use of $\bar{X}$ and $\bar{x}$. Further, $E(l_3|I) = \bar{y}$, as noted earlier. Hence, another estimator for $\bar{Y}$ that can be considered is

$$l_4 = l_3 + b^*(\bar{X} - \bar{x}) \tag{26}$$

Since $E(l_4|I) \doteq \bar{y} + b(\bar{X} - \bar{x}), V[E(l_4|I)] = (1-f)S_e^2/n$. A large sample approximation to the variance of l_4 is

$$V_4 \doteq \frac{(1-f)}{n} S_e^2 + \frac{W_2(k-1)}{n} S_{y2}^2(1-\rho_2^2) \quad (27)$$

This variance is clearly smaller than the variance of l_3 in (25). Further, from (19) and (27),

$$V_1 - V_4 = \frac{W_2(k-1)}{n}(\beta_2 - \beta)^2 S_{x2}^2 \quad (28)$$

Hence, the variance of l_4 is smaller than the variance of l_1, unless $\beta_2 = \beta$.

An estimator of V_4 is obtained by replacing S_e^2 by $\hat{S}_e^2$ given in the previous section and S_{y2}^2 by s_{em}^2.

6. OPTIMUM SAMPLE SIZES

The optimum value of k for each of the estimators can be obtained as described in Section 2. From this procedure, the optimum values of k for l_1 and l_2 are given by

$$k_1^2 = (t_1 - W_2)q \quad (29)$$

and

$$k_2^2 = (t_2 - W_2)q \quad (30)$$

where $t_1 = (S_e^2/S_{y2}^2)$ and $t_2 = S_y^2/S_{y2}^2(1-\rho_2^2) = t/(1-\rho_2^2)$. If $(S_y/S_{y2}) = (S_x/S_{x2})$, t_1 becomes equal to $t(1-\rho^2)/(1+\rho^2-2\rho\rho_2)$. We recall from Section 1 that $q = c_2/(c_0 + c_1 W_1)$ and $t = (S_y^2/S_{y2}^2)$.

Similarly, the optimum values of k for l_3 and l_4 are given by

$$k_3^2 = (t_3 - W_2)q \quad (31)$$

and

$$k_4^2 = (t_4 - W_2)q \quad (32)$$

where $t_3 = S_y^2/S_{y2}^2(1-\rho_2^2) = t/(1-\rho_2^2)$ and $t_4 = S_e^2/S_{y2}^2(1-\rho_2^2) = t(1-\rho^2)/(1-\rho_2^2)$.

With the above values of k, for a given cost the optimum size for the initial sample is obtained from (7). We note that for obtaining the above optimum values, no extra costs are considered for measuring the auxiliary variable.

The optimum value for n for a given value V for the variance is given in (6). Similarly, from (19), (21), (25), and (27), the optimum sizes of the initial sample for a given variance V of $(l_1 - l_4)$ are

$$n_{1\text{opt}} = \frac{N}{NV + S_e^2}[S_e^2 + W_2(k_1 - 1)S_{e2}^2] \tag{33}$$

$$n_{2\text{opt}} = \frac{N}{NV + S_e^2}[S_e^2 + W_2(k_2 - 1)S_{y2}^2] \tag{34}$$

Table 1. Variances of the Regression Estimators $(l_1 - l_4)$ and the Sample Mean $\bar{y}^*$ when $\rho_1 = .8$ and ρ_2 Takes Different Values. $N_1 = 600$, $N_2 = 400$, $N = 1000$, $n = 50$. See Section 7 and Figure 1.

		Subsampling Fraction (1/k)			
ρ_2	ρ	1/5	1/4	1/3	1/2
.4	.73	8.36	6.89	5.42	3.96
		7.09	5.94	4.79	3.64
		9.19	8.23	7.26	6.29
		6.36	5.34	4.42	3.45
.5	.76	7.38	6.10	4.82	3.53
		6.86	5.70	4.55	3.40
		8.78	7.91	7.05	6.19
		5.70	4.84	3.98	3.11
.6	.79	6.33	5.25	4.17	3.08
		6.61	5.45	4.31	3.15
		8.27	7.54	6.80	6.06
		4.95	4.21	3.48	2.74
.7	.82	5.20	4.34	3.48	2.61
		6.36	5.20	4.05	2.90
		7.67	7.09	6.50	5.91
		4.10	3.51	2.92	2.33
.8	.85	4.01	3.38	2.74	2.11
		6.09	4.94	3.79	2.63
		6.98	6.57	6.15	5.74
		3.14	2.73	2.31	1.90
$V(\bar{y}^*)$		9.93	8.78	7.62	6.47

$$n_{3\text{opt}} = \frac{N}{NV + S_y^2} [S_y^2 + W_2(k_3 - 1)S_{y2}^2(1 - \rho_2^2)] \tag{35}$$

and

$$n_{4\text{opt}} = \frac{N}{NV + S_e^2} [S_e^2 + W_2(k_4 - 1)S_{y2}^2(1 - \rho_2^2)] \tag{36}$$

7. ILLUSTRATIONS

For the sake of illustration, we have considered a population of $N = 1000$ units, with $N_1 = 600$ respondents and $N_2 = 400$ nonrespondents. The population values of the scores x and y on two tests for the respondents are $\bar{X}_1 = 75$, $\bar{Y}_1 = 90$, $S_{x1}^2 = 81$, $S_{y1}^2 = 121$, and $S_{xy1} = 79.2$. Corresponding figures for the nonrespondents are $\bar{X}_2 = 60$, $\bar{Y}_2 = 65$, $S_{x2}^2 = 196$, and $S_{y2}^2 = 144$. We have considered the value of ρ_2 to range from .4 to .8, with an interval of .1. We have considered $n = 50$ and varied k from 2 to 5 with unit intervals. Thus, the subsampling fraction ranges from 20 to 50 percent. The variances of the different estimators are shown in Table 1 and Figure 1 when $\rho_1 = .8$ and ρ_2 ranges from .4 to .8, and in Table 2 and

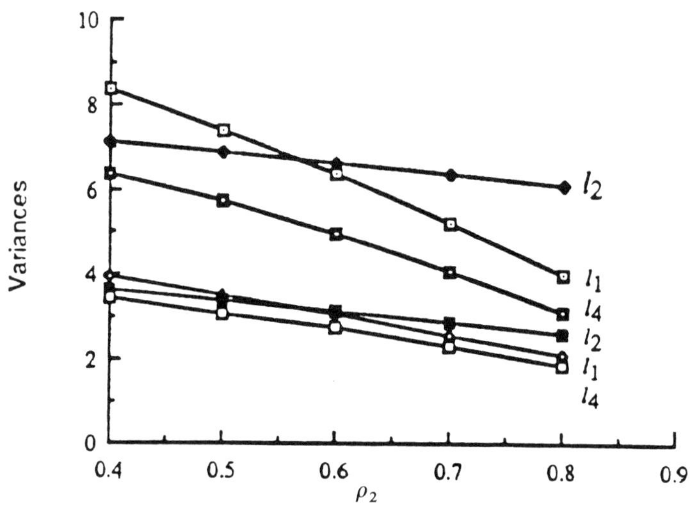

Figure 1. Variances of l_1, l_2, and l_4. Top figures refer to $k = .2$ and bottom figures refer to $k = .5$.

Table 2. Variances of $(l_1 - l_4)$ when $\rho_2 = .8$ and ρ_1 Takes Different Values. $N_1 = 600$, $N_2 = 400$, $N = 1000$, $n = 50$. See Section 7 and Figure 2.

ρ_1	ρ	Subsampling Fraction $(1/k)$ 1/5	1/4	1/3	1/2
.4	.74	4.40	3.89	3.39	2.88
		6.98	5.83	4.68	3.53
		6.98	6.57	6.15	5.74
		4.04	3.62	3.21	2.79
.5	.77	4.29	3.76	3.23	2.70
		6.77	5.62	4.47	3.32
		6.98	6.57	6.15	5.74
		3.82	3.41	2.99	2.58
.6	.80	4.19	3.63	3.07	2.51
		6.55	5.40	4.25	3.10
		6.98	6.57	6.15	5.74
		3.60	3.19	2.77	2.36
.7	.82	4.10	3.50	2.91	2.31
		6.32	5.17	4.02	2.87
		6.98	6.57	6.15	5.74
		3.38	2.96	2.55	2.13
.8	.85	4.01	3.38	2.74	2.11
		6.09	4.94	3.79	2.63
		6.98	6.57	6.15	5.74
		3.14	2.73	2.31	1.90
$V(\bar{y}^*)$		9.93	8.78	7.62	6.47

Figure 2 when $\rho_2 = .8$ and ρ_1 ranges from .4 to .8.

We note that

$$(N-1)S_x^2 = (N_1 - 1)S_{x1}^2 + (N_2 - 1)S_{x2}^2 + NW_1W_2(\bar{X}_1 - \bar{X}_2)^2 \quad (37)$$

$$(N-1)S_y^2 = (N_1 - 1)S_{y1}^2 + (N_2 - 1)S_{y2}^2 + NW_1W_2(\bar{Y}_1 - \bar{Y}_2)^2 \quad (38)$$

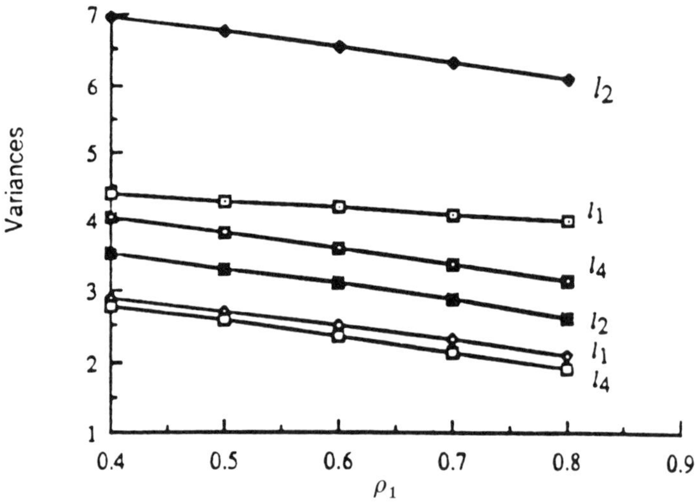

Figure 2. Variances of l_1, l_2, and l_4. Top figures refer to $k = .2$ and bottom figures refer to $k = .5$.

and

$$(N-1)S_{xy} = (N_1 - 1)S_{xy1} + (N_2 - 1)S_{xy2} + NW_1W_2 \times (\bar{X}_1 - \bar{X}_2)(\bar{Y}_1 - \bar{Y}_2) \tag{39}$$

These formulas are useful for finding the variances and covariances for the entire population from the corresponding values for the two strata. The population correlation coefficient ρ in Tables 1 and 2 is obtained from these expressions.

The optimum value of k for $\bar{y}^*$ is presented in (5), and the optimum value of n for a given cost is presented in (7). The optimum values of k for $(l_1 - l_4)$ are presented in (29) through (32). As mentioned in Section 6, the optimum values of n for $(l_1 - l_4)$ for a given cost are obtained from (7) with the appropriate value of k. For the sake of illustration, we have considered $c_0 = 1$, $c_1 = 5$, $c_2 = 25$, and a total available amount of $C = 1200$. If there is no nonresponse, the size of the sample with these costs is equal to $n = C/(c_0 + c_1) = 200$. The optimum values of n and k for each of the estimators and the minimum variances are presented in Table 3 and

Table 3. Optimum Values of the Initial Samples Sizes, Subsampling Fractions, and the Minimum Variances of $(l_1 - l_4)$ and $\bar{y}^*$ when $\rho_1 = .8$ and ρ_2 Takes Different Values. $C = 1200$, $c_0 = 1$, $c_1 = 5$, $c_2 = 25$. See Section 7 and Figure 3.

ρ_2	Sample Size	Subsampling Fraction (%)	Minimum Variance
.4	108	72	1.43
	125	56	1.36
	174	29	2.21
	136	48	1.30
.5	110	69	1.28
	118	62	1.26
	179	27	2.14
	136	48	1.17
.6	114	65	1.13
	110	70	1.14
	186	25	2.04
	139	46	1.02
.7	121	60	0.97
	98	82	1.02
	195	22	1.91
	144	44	0.87
$V(\bar{y}^*)$	166	32	2.33

Figure 3 when $\rho_1 = .8$ and ρ_2 ranges from .4 to .7, and in Table 4 and Figure 4 when $\rho_2 = .8$ and ρ_1 ranges from .4 to .7.

Since l_3 has relatively large variances, it is not shown in Figures 1 through 4.

8. CONCLUSIONS

Comparisons of the variances of $(l_1 - l_4)$ presented in Sections 4 and 5 lead to the following conclusions.

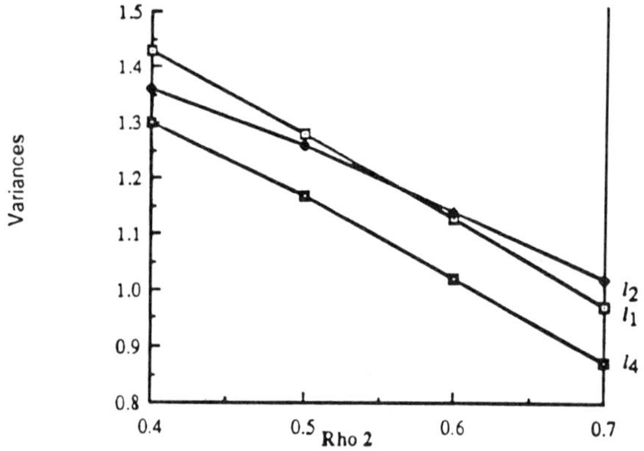

Figure 3. Minimum variances of l_1, l_2, and l_4, when rho 1 is .8.

Table 4. Optimum Values of the Initial Sample Sizes, Subsampling Fractions, and the Minimum Variances of $(l_1 - l_4)$ and $\bar{y}^*$ when $\rho_2 = .8$ and ρ_1 Takes Different Values. $C = 1200$, $c_0 = 1$, $c_1 = 5$, $c_2 = 25$. See Section 7 and Figure 4.

ρ_2	Sample Size	Subsampling Fraction (%)	Minimum Variance
.4	167	32	1.04
	122	58	1.31
	207	18	1.74
	176	28	0.98
.5	160	35	0.98
	115	64	1.22
	207	18	1.74
	172	30	0.91
.6	152	39	0.93
	107	72	1.12
	207	18	1.74
	167	32	0.85
.7	142	44	0.87
	97	84	1.00
	207	18	1.74
	161	35	0.78
$V(\bar{y}^*)$	166	32	2.33

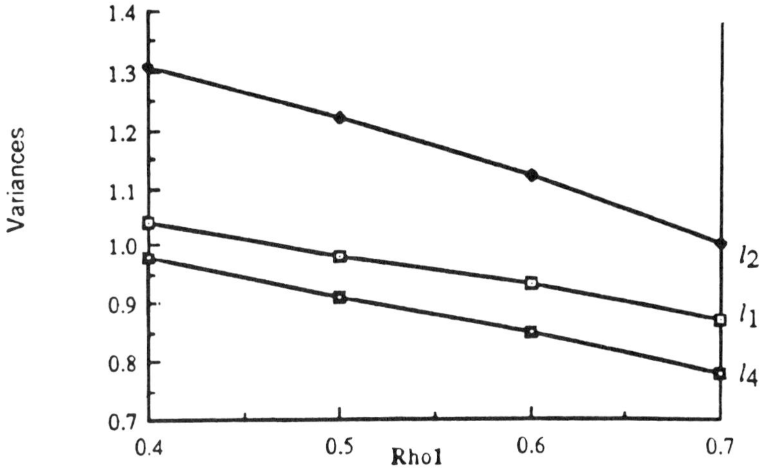

Figure 4. Minimum variances of l_1, l_2, and l_4, when rho 2 is .8.

1. The estimator l_4, in general, has smaller variance than the remaining estimators, with l_1 coming next. The difference between l_1 and l_4 becomes negligible if the subsampling fraction is more than 50 percent. In this case, l_1 may be preferred to l_4, since it has a relatively simple expression. See Tables 1 through 4 and Figures 1 through 4.
2. Although l_2 uses the information on x from all the n sampled units, its variance would be smaller than that of l_1 only if ρ_2 is small.
3. For all the estimators, relatively larger values of ρ_2 than ρ_1 result in smaller variances. Compare the variances in Tables 1 and 2 or in Tables 3 and 4. Compare Figures 1 and 2 or Figures 3 and 4.
4. As expected, variances of $(l_1 - l_4)$ are smaller than the variance of $\bar{y}^*$. They also decrease with increasing subsampling fractions and increasing correlation between x and y.

REFERENCES

Cochran, W. G. (1977). *Sampling Techniques*, New York, John Wiley & Sons.

Hansen, M. H. and Hurwitz, W. N. (1946). The Problem of Nonresponse in Sample Surveys, *J of the American Statistical Association*, *41*: 517–529.

Rao, J. N. K. (1973). On Double Sampling for Stratification and Analytical Surveys, *Biometrika, 60*: 125–133.

Rao, P. S. R. S. (1983). Randomization Approach, *In Incomplete Data in Sample Surveys,* Vol. 2; Eds. W. G. Madow, I. Olkin, and D. B. Rubin, New York, Academic Press, pp. 33–44.

Rao, P. S. R. S. (1986). Ratio Estimation with Subsampling the Nonrespondents, *Survey Methodology, 12*(*2*): 217–230.

Rao, P. S. R. S. (1987). *Ratio and Regression Estimates with Subsampling the Nonrespondents.* Paper presented at a special contributed session of the International Statistical Association Meetings, September 2–16, 1987, Tokyo, Japan.

Srinath, K. P. (1971). Multiphase Sampling in Nonresponse Problems, *J of the American Statistical Association, 16*: 583–586.

13

SUDAAN: A Comprehensive Package for Survey Data Analysis

LISA M. LaVANGE*, **BABUBHAI V. SHAH**, **BETH G. BARNWELL**, and **JOYCE F. KILLINGER** Research Triangle Institute, Research Triangle Park, North Carolina

The development of a survey data analysis software package that is flexible enough to meet the needs of survey statisticians, as well as research scientists not necessarily schooled in survey sampling, is underway at the Research Triangle Institute. The system was designed to provide significant enhancements to RTI's existing software package, SUDAAN. The new SUDAAN system is written in the C language and provides greater efficiency and portability. The software development strategy consists of developing and debugging stand-alone statistical procedures on an IBM PC with expanded capacity. The procedures are then implemented in two additional computing environments, VAX/VMS and IBM/OS. Variance estimation options and data analysis techniques not previously available in the RTI software package are among the many modifications incorporated into the new SUDAAN system. The purpose of this paper is to describe the design and development of this system. *Keywords: variance estimation, user interface, C language.*

1. INTRODUCTION

Scientists at the Research Triangle Institute have been involved in designing, developing, and maintaining software systems for the analysis

**Current affiliation*: Frank Parker Graham Child Development Center, University of North Carolina, Chapel Hill, North Carolina

of survey data for the past 17 years. Recently, RTI has been under contract to the Public Health Service (PHS) to develop a comprehensive software package that meets the needs of statistical analysts at the National Center for Health Statistics (NCHS) and PHS. In particular, this package must be flexible enough to handle most of the statistical designs used by these agencies. For this purpose, we have embarked on the task of developing a system that incorporates many of the features of RTI's existing survey data analysis system but also includes significant enhancements. The purpose of this paper is to describe in detail the design and development of this comprehensive software package. The immediate objective of the SUDAAN system design was to develop a series of procedures capable of performing statistical data analysis for complex sample surveys. The ultimate objective was to have a system that could aid in the design and evaluation of new statistical techniques.

RTI's existing software system consists of a series of procedures that produce statistical analyses in the form of weighted cross-tabulations, generalized ratio estimators, and linear and logistic regressions. All of the procedures were constructed using a unified set of FORTRAN subroutines that execute in the SAS computing environment on an IBM mainframe. In the course of developing a new system, we went through the process of rethinking our current design. Four goals emerged:

Portability
Reliability/numerical accuracy
Computational efficiency
Ease of modification/enhancement

The system design for the comprehensive survey data analysis software, SUDAAN, evolved with these goals in mind. The first goal is being met by writing the new software entirely in the C programming language and testing simultaneously on several systems. Compilers for the C language are available on a variety of operating systems. The initial programs are currently executing under the TURBO C compiler on an IBM personal computer, the SVC compiler on a 68020, and under the UNIX operating system on a GOULD computer. Versions are also being tested on the VAX/VMS and IBM/VS.

The second goal is being met by building checks into the system to assure that definitions of objects are meaningful and computations are feasible. We have implemented numerically stable procedures for the accumulation of sums of squares and cross products and for matrix operations, including matrix inversion.

For achieving the third goal of improved efficiency, special care has been taken in the design to isolate tasks that are repeated numerous times, such as a numerical calculation on each variable of each record. These sections of code are optimized to the extent possible without using assembly language. Some of the optimizing algorithms include processing of sparse matrices and vectors, replacement of multiplication by table lookup with addition, elimination of repetitive (in the loop) "if" tests by constructing separate do loops, and many others.

An important product of the SUDAAN design effort has been the creation of a high-level statistical programming language. The SUDAAN language has been developed to provide the statistician/programmer with easy access to the software package, thus accomplishing the fourth goal listed above. Statistical procedures are currently being written in the SUDAAN language. Once they are finalized, some parts may be converted to C programs to maximize computing efficiency.

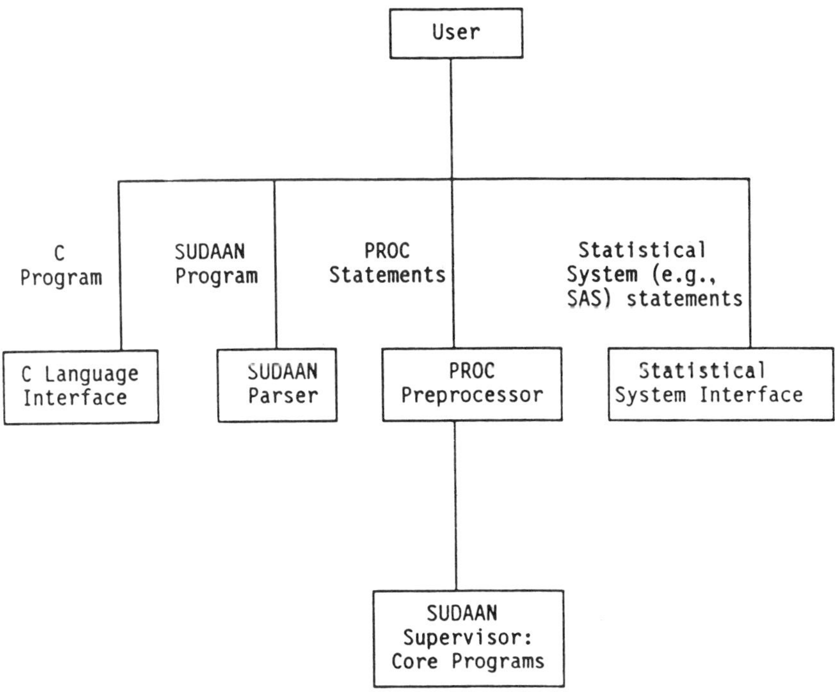

Figure 1. User access to the SUDAAN system.

The SUDAAN design allows for four types of user interface as depicted in Figure 1. The general user will be able to access the system through user friendly procedures. Ultimately, the procedures will be available through a statistical system, such as SAS, but the current design consists of stand-alone procedures that interface with a standard ASCII file format. The more sophisticated user can write SUDAAN programs for computations not available in the procedure library. This access feature is particularly well suited for testing modifications to existing procedures, such as the calculation of an alternative test statistic. A C programmer can access the system directly through C callable functions.

The key features of the SUDAAN system are described in detail in the following chapters. A brief overview of the software package for survey data analysis currently marketed by RTI is given in the following section. A description of the procedures being offered through the new software package, with emphasis on how the two packages differ, is also provided. The SUDAAN system design is the next topic. The SUDAAN language is illustrated with programming examples. The advantages and disadvantages of the SUDAAN design are the topic of the final section.

2. RTI SURVEY DATA ANALYSIS PROCEDURES

2.1 Background

The collaboration of statisticians and computer scientists at RTI, beginning in 1971 (e.g., Folsom, Bayless, and Shah, 1971) and continuing today, has resulted in a series of programs for survey data analysis that reflects state-of-the-art methodology in this field. All of RTI's analysis packages employ the Taylor series or Delta method of variance estimation for complex sample survey designs (Folsom, 1974). Each program currently marketed is designed to run within the SAS framework.

Early work involved a variance component procedure developed to produce estimates of variance components associated with the strata and stages of an unequally weighted, stratified cluster sample. An updated version, VCMPNLS (Shah, 1979) became available as a SAS procedure in 1979. Weighted estimates of means, totals, and proportions and their standard errors were first provided by the program STDERR in 1974. A later version, SESUDAAN (Shah, 1981), enables the user to test for differences between domain estimates in a two-way cross-classification and to produce estimates standardized to a user-specified population distribution. Modifications to SESUDAAN in 1980 and 1981 resulted in two procedures, RATIOEST (Shah, 1981) and RTIFREQS (Shah, 1982). RATIOEST produces estimates of ratios of two variables. RTIFREQS calculates weighted frequency distributions and estimates of row, column, and overall percentages for user-specified cross-tabulations. A further modification to RATIOEST that allows the user to input a sample unit level file and an event level file (e.g., medical visits with multiple records per subject) resulted in the RATIO2 procedure (Wheeless and Shah, 1982).

RTI's survey regression package, SURREGR (Holt, 1977), provides estimates of linear regression coefficients, their estimated variance–covariance matrix, and tests of hypotheses concerning the coefficients that are appropriate for a complex sample design. Logistic regression capabilities became available with the development of RTILOGIT (Shah et al., 1984). This procedure runs in conjunction with SAS PROC LOGIST and produces pseudo–maximum likelihood estimates of the parameters and tests of hypotheses about these parameters.

RTI is currently marketing the following procedures for general use:

SESUDAAN
RATIOEST
RTIFREQS
SURREGR
RTILOGIT

As mentioned earlier, variance estimation for all of these procedures is performed using the Taylor series linearization method. The RTIFREQS, SURREGR, and RTILOGIT procedures assume with-replacement sampling, thereby enabling a first-stage variance formula to be used. This approach is generally thought to be a conservative approximation for variances associated with more complicated designs. The SESUDAAN and RATIOEST procedures allow the user to specify with- or without-replacement sampling at each stage of a multistage design. Test statistics produced in SESUDAAN, SURREGR, and RTILOGIT are based on Hotelling's T^2 type statistic which is assumed to have a transformed F distribution in repeated samples (Shah et al., 1977).

2.2 New System Procedures

The SUDAAN system design consists of two phases of software development, each resulting in a series of statistical procedures. The Phase I procedures are primarily concerned with descriptive data analysis, including cross-tabulations, generalized ratio estimation, and quantile estimation. The Phase II design, currently in the planning stages, will incorporate several analytical procedures, including linear and log-linear modeling and survivorship analysis of survey data.

Several features are available under the new SUDAAN system that were not previously available with RTI software. Key among these are multiple variance options. Sampling variances are computed for one of three design options invoked by the user. These are:

With-replacement sampling at the first stage
Simple random sampling at each stage (with or without replacement)
Without-replacement, unequal probability sampling at the first stage and simple random sampling (with or without replacement) at subsequent stages

The second option requires the input of stratum specific population and sample counts for each stage. In addition to these counts, the third option requires the input of joint probabilities of selection for each pair of primary sampling units within a stratum. These quantities are used to produce Yates–Grundy–Sen variance estimates for this design option. The three variance options cover a wide range of sample designs including sampling with probabilities proportional to size at the first stage. Unequal probabilities of selection at a subsequent stage can be handled by forming pseudostrata at that stage since probabilities of selection can vary from stratum to stratum. In addition to these variance options, the user may request that the variance–covariance matrix for estimates within a table be computed for all procedures allowing for cross-tabulations. Estimates of variance–covariance matrices were previously available only in the SURREGR and RTILOGIT procedures.

Phase I software consists of three procedures: CROSSTAB, DESCRIPT, and RATIO. The CROSSTAB procedure is the RTIFREQS equivalent in the new SUDAAN system. In addition to computing weighted frequency and percentage distributions, chi-square tests of independence for a two-way contingency table are produced. The tests are computed using a Wald statistic according to methods advocated by Koch et al. (1975). During phase II, the Rao–Scott Satterthwaite corrected test statistic (Thomas and Rao, 1984, 1985) will be computed in addition to the Wald statistic for goodness-of-fit tests. This statistic has been shown by Thomas and Rao to have the best characteristics with respect to both power and significance level when compared with other statistics correcting for the liberalness of the Wald test.

The DESCRIPT procedure computes estimates of means, totals, proportions, geometric means, quantiles, and their sampling errors. Estimates can be requested for the subdomains of user-specified cross-tabulations. Options are also available to the user for standardized and/or poststratified means and proportions and their sampling errors. The standardizing and/or poststratifying variables and their distributions are required as input.

RTI has recently completed a simulation study comparing two methods for estimating the variances of quantiles for sample survey data (Wheeless et al., 1987). The two methods considered were (1) the one proposed by Francisco and Fuller (1986) and currently used in the PC CARP software package and (2) a direct linearization method proposed by RTI. The simulation also compared two methods of estimating quantiles themselves. Based on the results of this study, the direct linearization method is currently being programmed for use in the DESCRIPT procedure.

The user can request estimates of linear contrasts among the domain estimates produced by DESCRIPT. Four types of contrast specifications are available. The user can specify a general linear contrast, all possible pairs of differences among levels of a domain variable, the differences between levels of a domain variable and the marginal value, and polymonial effects for an ordinal domain variable.

The RATIO procedure computes the estimates of generalized ratios for survey data. The user can specify numerator and denominator variables that are continuous or categorical. In the latter case, levels of the variables for which ratios are to be estimated are required. This procedure, as the DESCRIPT procedure, can be executed in one of three modes: basic estimation, standardized estimation, or poststratified estimation.

The current plan for phase II software development consists of four statistical procedures written in the new SUDAAN framework. The first of these is a general categorical data analysis procedure that will offer log-linear modeling capabilities for survey data analysis. This procedure will be a survey data analogue to the SAS procedure CATMOD. Procedures two and three will be linear and logistic regression procedures, developed based on the existing procedures SURREGR and LOGIST. Each of these will be reprogrammed in the new SUDAAN system and various enhancements added. Enhancements to the linear regression procedure are being planned for stochastic regression coefficients model analysis of survey data. Software has been developed at RTI under a contract to the National

Institute for Child Health and Human Development for this problem (LaVange et al., 1988) that will be incorporated into the SUDAAN system. Fourth, a general procedure for survival data analysis is planned that will allow the analyst to perform Kaplan–Meir estimation and Cox proportional hazard model estimation using complex sample survey data.

3. SUDAAN SYSTEM DESIGN

The immediate objective of the SUDAAN system design was to develop a series of procedures capable of performing statistical data analysis for complex sample surveys. The ultimate objective was to have a system that could aid in the design and evaluation of new statistical techniques. The system design for SUDAAN consists of three layers:

1. PROCs or procedures similar to SAS or BMDP procedures
2. SUDAAN data structures and functions operating on these data structures
3. A system level interpreter for layers (1) and (2)

These layers correspond to different views of the system. SUDAAN may be perceived quite differently by different groups of users. In this section we discuss these views and describe in detail the major components of the SUDAAN system.

3.1 System Views

A substantive researcher or data analyst will view SUDAAN as a collection of statistical procedures for applying alternative statistical techniques for survey data analysis. A statistician may view it as a convenient tool for evaluating new statistical formulas or analysis techniques. SUDAAN allows for easy conversion of formulas into executable statements, thereby enabling new procedures or enhancements to existing procedures to be programmed quickly and efficiently.

A system designer may view it as a compiler, written in C, for a very high level language in order to specify statistical analysis tasks. For the computer scientist, the development effort consisted of designing an optimizing compiler or interpreter with the potential for global optimization of all available resources. The SUDAAN program defines only what is to be done, not how. The system is free to select any of the possible sequences of operations in order to achieve the user-defined result.

The PROC interface is in SUDAAN for the convenience of the user who repeatedly applies the same procedures to different datasets or different variables within the same dataset. The systems programmer's view represents one of the many possible compiler designs for the implementation of SUDAAN,

However, the primary force behind the SUDAAN design is the statistician's view, namely, a tool for translating mathematical formulas into a computer program capable of producing numerical results for a given data configuration. The statistician need not specify how a particular estimator is to be calculated. Rather his focus is on what is to be calculated. For the statistician, SUDAAN is a "nonprocedural" language. Given the definition of a statistical quantity in terms of a mathematical formula, the system determines the computational algorithm. Users have no knowledge of the representation of the data structures in memory or on an input/output device and no control of the program flow. In this sense the SUDAAN language is viewed as definitional and not procedural.

The basic elements of the system are as follows:

Statistical data objects and abstract data structures
Functions of and operations to data structures
Functions to input, output, and print data structures

These are described and illustrated in the following subsections.

3.2 Statistical Data Structures

An example of a statistical data structure is the dataset resulting from a national household survey. Such a dataset might consist of records containing information on a sample of individuals, each selected from a sample of households within enumeration districts, counties, and states. The actual dataset contains a record for each individual, consisting of various attributes and item responses for that individual and household. Each record is identified by several hierarchical identifiers corresponding to the stages of sample selection: state, county, enumeration district, household, and individual. An example of a different type of data structure is a summary dataset of county-specific population counts of individuals by characteristics, such as race and sex. Such a dataset is a collection of two-dimensional tables or matrices with the two hierarchical identifiers state and county.

In SUDAAN, statistical data structures are defined as collections of multidimensional arrays with nested hierarchical identifiers. These data structures are referred to as balanced array trees (BATs). Each of the datasets described above is an example of a BAT, the first with five nested identifiers and the second with only two. The SUDAAN language also consists of a series of functions and mathematical operators that are used to input and output BATs and to define BATs as functions of BATs and other imput objects. The SUDAAN language itself is a mathematically closed set of BATs with respect to these functions. Applying a function to a BAT yields another BAT. The level of nested identifiers may change, but one or more BATs always result.

3.3 SUDAAN Program Examples

In order to illustrate the SUDAAN language concepts, consider the problem of computing the between primary sampling unit (PSU), within stratum mean square for an estimated population total. Let $y(hijk)$ denote the observed values of the variable y collected from a nested sample design for the kth individual in the jth household, ith PSU, and hth stratum. The filename "EXAMPLE1" refers to an ASCII file with the nested sample identifiers included as variables one to four on the file. The variable y is the fifth variable on the file. The input file is stored in a standard file format with a codebook describing the variables and associated levels. Variable labels or formats are automatically read in and used for printing. In the following statements, X refers to the BAT consisting of these data values. The program uses the summation function, SIGMA, to compute totals of y at various levels. The first argument of SIGMA refers to the BAT containing data values to be summed, while the second argument refers to the nesting level up to which the summing takes place. This argument equals the number of nested identifiers in the resulting BAT that contains the sums. Comments, delimited by the strings "/*" and "*/", further clarify the logic of the following program statements.

```
RECORDS=FICSFILE("EXAMPLE1",
          VECTOR(1,2,3,4));              /*RECORDS is a BAT containing the
                                           input dataset*/
X=SELECT(RECORDS,5);                     /*X is a BAT of y values and id's */
PSULEVEL=2;
XSUM2=SIGMA(X,PSULEVEL);                 /*Sum X over each PSU */
MH=SIGMA(CONSTANT(XSUM2,1),1);           /*Compute the number of PSUs
                                           per stratum by defining a vector
                                           of one's for each PSU in XSUM2*/
MHMIN1=MH-1;
STRATLEV=1;
XSUM1=SIGMA(XSUM2,STRATLEV);             /*Sum X over each stratum */
```

```
XSQR1=SIGMA(XSUM2*XSUM2,
            STRATLEV);                  /*Stratum sum of squares of
                                          PSU totals */
XSS0=SIGMA((MH*XSQR1 -
      XSUM1*XSUM1)/MHMIN1,0);           /*Compute sum of squared deviations
                                          over all strata */
MSE=SQRT(XSS0);
PRTABS(MSE);                            /*Print the result */
```

Since the SUDAAN language deals with symbolic definition only, the above statements can easily be modified to obtain the mean square errors associated with more than one variable. Replacing the second statement with the following results in the calculation of mean square errors for variables statement 5, 6, 9, and 10 in this example.

```
X = SELECT(RECORDS, VECTOR(5,6,9,10));
```

Next, consider the problem of computing mean square errors for estimated population counts corresponding to the cells of a multidimensional table. In the following example statements, TABLES refers to a BAT consisting of a two-way and a three-way table. The first is the cross-classification of variables 5 and 6 on the input file and the second is the cross-classification of variables 5, 9, and 10. The EFFECTS function converts categorical variables with specified numbers of levels into dummy (0, 1) vectors. The CROSSP function defines the tables as cross products of these effect vectors. SIGMAWG returns two arguments, the unweighted and weighted sums for each cell in the tables. Once XSUM2 is defined, the program statements in the preceding example follow to compute the mean square errors. The required statements are:

```
RECORDS=FICSFILE("EXAMPLE1", VECTOR(1,2,3,4));
SUBGROUPS=VECTOR(5,6,9,10);     /*Variables defining subgroups */
LEVELS=VECTOR(7,2,2,2);         /*Associated variable levels */
OPTIONS=VECTOR(0,0,0,0);        /*Four options are off */
EFFVECTORS=EFFECTS(RECORDS,SUBGROUPS,LEVELS,OPTIONS);
TAB1=VECTOR(1,2);               /*Variable 5 by variable 6 */
TAB2=VECTOR(1,3,4);             /*Variable 5 by var 9 by var 10 */
TABVEC=VECTOR(TAB1,TAB2);
TABLES=CROSSP(EFFVECTORS,TABVEC);
WEIGHT=SELECT(RECORDS,11)       /*Variable 11 is the weight */
NWSUM=SIGMAWG(TABLES,PSULEVEL,WEIGHT);
XSUM2=NWSUM(2);
```

Additional statements can be added to these examples to compute mean square errors for estimated percentages for each table. These require additional SUDAAN functions and are illustrated in the complete program example given in Appendix A. This program illustrates variance estimation for the CROSSTAB procedure assuming stratified simple random sampling with replacement.

3.4 SUDAAN Procedures

The task of developing new procedures in SUDAAN is fairly straightforward. The process involves writing a SUDAAN program and associating objects in the program with a well-defined set of key words. The user input to a PROC consists of a command level syntax, such as that used in IBM/TSO or SAS. The PROC processor interprets the user input and generates the necessary statements for the SUDAAN language parser. In short, writing a SUDAAN procedure is equivalent to defining a preprocessor for a SUDAAN program.

The major advantage of this approach is that the user has a SUDAAN program available for each procedure that serves as detailed documentation. The user can see the translation of input to output and the series of calculations that are performed once the procedure has been called. Modifications to the procedure can be made by modifying the accompanying SUDAAN program without writing a completely new procedure. The source for all of the SUDAAN procedures will be available to the user. An example is the source for the CROSSTAB procedure, also provided in Appendix A.

3.5 System Components

The system is composed of six major components:

Lexical analyzer
Syntax analyzer
Semantic analyzer
Executing supervisor
Function definitions
PROC processor

The overall system flow is depicted in Figure 2. As mentioned earlier, the entire system is written in the C programming language. The user-supplied program is first passed to a lexical analyzer and a symbol table is output. This table is processed by the syntax analyzer into a list of executable statements. The semantic analyzer checks for valid function types, compatible arguments, and meaningful interpretation of the program statements. Error messages are supplied if necessary at this step. An intermediate output step is available to store the parsed program if so desired. An optimizing scheduler rearranges the order of execution of the program statements for optimal resource utilization. The executing supervisor then

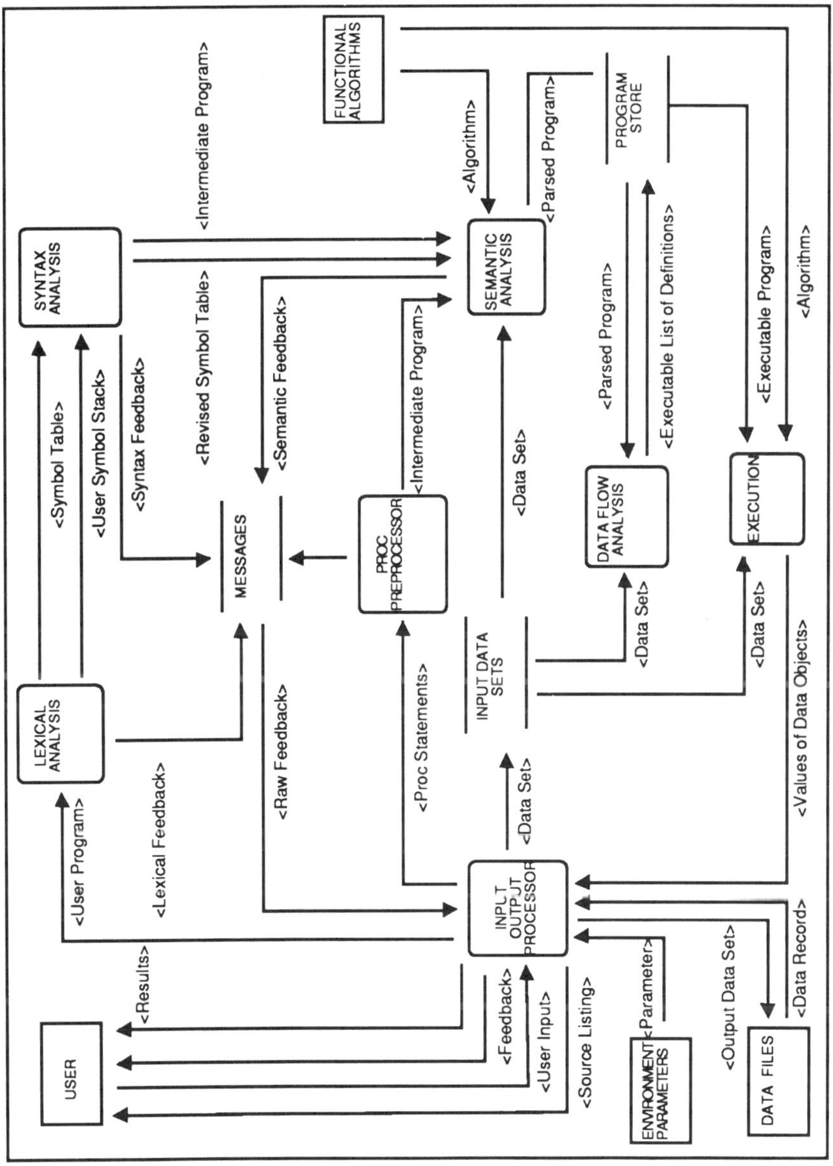

Figure 2. SUDAAN system design.

produces the program results. The input/output processer provides the appropriate user interface to the system. This may include a procedure interface that transforms procedural statements into a series of arguments for the SUDAAN program.

4. CONCLUDING REMARKS

The unique feature of SUDAAN is the entity defined as a BAT. Mathematically, a BAT represents a finite sequence of multidimensional arrays. The set of BATs represents a closed set with respect to binary operations and functions whose arguments are BATs. The functional syntax is "natural" for a mathematical statistician who can use the SUDAAN language to define the BATs of interest. A SUDAAN user will have the SUDAAN source code available as part of the documentation. By referring to the SUDAAN program for a statistical procedure, the analyst will be able to determine exactly what underlying mathematical formulas are employed. When statistical research indicates new or improved formulas for a given analysis, the formulas can be readily translated into the SUDAAN language to yield a working computer program.

In writing a SUDAAN program, the user only defines the computational objects through function cells. The user does not need to specify how the calculations will be carried out. Increased reliability is achieved through checks built into the system to assure that the definitions are meaningful and that the objects required as output can be generated. The system is free to select the computational algorithm that is optimal given the dataset and available computing resources. The system also provides tremendous flexibility with respect to input and output. All input/output objects are defined as BATs. Any intermediate BAT can be saved or displayed at any time in the program. The use of BATs, with the above stated advantages, has some restrictions and limitations. The user is limited to the BAT functions currently implemented, and all computations must be represented as BATs or functions thereof.

The SUDAAN programmer is confined to the available vocabulary of functions. Computations that require new functions of BATs may have to wait until such functions are implemented in SUDAAN. Over time, as more and more functions are implemented, this limitation will not be critical. The modular design of the system permits new functions to be added to the system without recompiling the entire system. Only the new function and its interface need to be compiled. It is also possible to test the new function independently before incorporating it into the system, thereby greatly facilitating the addition of functions to the system library on an as-needed basis.

Computations that cannot be represented as BATs will not be feasible in SUDAAN. The user must define all computations as functions of input BATs or data objects. The SUDAAN system is therefore not for general programming use, but should prove reasonably adequate for statistical analysis, and, in particular, survey data analysis.

A major portion of the overall software development effort at RTI has been channeled into the SUDAAN system design. While this strategy has slowed the development of the statistical procedures initially, we are anticipating a tremendous savings in the long run due to the ease with which each procedure can be designed, implemented, and debugged. The end result of this effort will be not only a user-friendly software package that runs efficiently but also a valuable aid to the statistician for enhancing existing procedures and planning for additional procedures as new methodology becomes available.

Acknowledgments

This work was partially supported by the Public Health Service through the National Center for Health Statistics under research contract 282–86–0107.

REFERENCES

Cox, D. R. (1972). Regression models and life tables (with discussion), *Journal of the Royal Statistical Society B34*:187–220.

Folsom, R. E., Bayless, D. L., and Shah, B. V. (1971). Jackknifing for Variance components in complex sample survey designs. *Proceedings of the American Statistical Association* 36–39.

Folsom, R. E. (1974). National assessment approach to sampling error estimation. *Sampling Error Monograph.* Prepared for the National Assessment of Educational Progress.

Francisco, C. A., and Fuller, W. A. (1986). Estimation of the Distribution Function with Data from a Complex Survey. Presented at the American Statistical Association Meetings, Chicago, Illinois.

Holt, M. M. (1977). *SURREGR: Standard Errors of Regression Coefficients from Sample Survey Data.* Research Triangle Institute, Research Triangle Park, NC.

Kaplan, E. L., and Meier, P. (1958). Nonparametric estimation from incomplete observations. *Journal of the American Statistical Association 53*:457–481.

Koch, G. G., Freeman, D. H., Jr., and Freeman, J. L. (1975). Strategies in the multivariate analysis of data from complex surveys. *Inter. Stat. Rev. 43*:59–78.

LaVange, L. M., Pfeffermann, D., Shah, B. V., and Gabel, T. J. (1988). The Analysis of Relationships Between Growth and Dietary Intake Using the NHANES Dataset. Final Project Report. Research Triangle Institute, Research Triangle Park, NC.

Shah, B. V., Holt, M. M., and Folsom, R. E. (1977). Inference About Regression Models from Sample Survey Data. *Bulletin of the International Statistical Institute* 41: 43–57.

Shah, B. V. (1981). *RATIOEST: Standard Errors Program for Computing Ratio Estimates from Sample Survey Data.* Research Triangle Institute, Research Triangle Park, NC.

Shah, B. V. (1982). *RTIFREQS: Program to Compute Weighted Frequencies, Percentages, and Their Standard Errors.* Research Triangle Institute, Research Triangle Park, NC.

Shah, B. V. (1981). *SESUDAAN: Standard Errors Program for Computing Standardized Rates from Sample Survey Data.* Research Triangle Institute, Research Triangle Park, NC.

Shah, B. V. (1978). SUDANN: Survey Data Analysis Software. Presented at the Joint Statistical Meetings of the American Statistical Association, San Diego.

Shah, B. V. (1979). *VCMPNLS: Program to Compute Variance Components.* Research Triangle Institute, Research Triangle Park, NC.

Shah, B. V., Folsom, R. E. Harrell, F. E., and Dillard, C. N. (1984). RTILOGIT: Survey Data Analysis Software for Logistic Regression, Final Report, Work Assignment 74, Research Triangle Institute, Research Triangle Park, NC.

Thomas, D. R., and Rao, J. N. K. (1984). A Monte Carlo study of exact levels of goodness-of-fit statistics under cluster sampling. *Proceedings of the Section on Survey Research Methods*, American Statistical Association, Washington, DC.

Thomas, D. R., and Rao, J. N. K. (1985). On the power of some goodness-of-fit tests under cluster sampling. *Proceedings of the Section on*

Survey Research Methods, American Statistical Association, Washington, DC.

Wheeless, S., and Shah, B. V. (1982). RATIO2 and RATIO3: Computing Ratio Estimates with Two Data Files. Research Triangle Institute, Research Triangle Park, NC.

Wheeless, S. C., Shah, B. V., and LaVange, L. M. (1987). Results of a Simulation for Comparing Two Methods for Estimating Quantiles and their Variances. Project Report. Research Triangle Institute, Research Triangle Park, NC.

APPENDIX A: SUDAAN PROGRAM EXAMPLE

```
/* Example of a call to PROC CROSSTAB */

PROC CROSSTAB DATA=test120 DESIGN=wrs;
NEST stratum psu segno sampno;
SUBGROUP agecat race dsmsa dpovind;
LEVEL 7 2 2 2;
WEIGHT LWT;
TABLES race*dpovind agecat*dpovind agecat*dsmsa*dpovind;
```

CROSSTAB: SUDAAN Program Statements

```
'include "constant.def"             /* file that defines constants */
filename="test120";                 /* input data file */
nestvar=vector(1,2,3,4);            /* position of stratum, psu, segment and
                                       person id on the input record. */
wgtvar=5;                           /* weight variable position on input data
                                       record */
desnam="WRS";                       /* design option equals with replace-
                                       ment sampling */
psulev=2;                           /* position of PSU variable on NEST
                                       statement vector */
strlev=1;                           /* position of stratum variable on NEST
                                       statement vector */
subgroups=vector(8,9,10,11);        /* positions of AGE, RACE, SMSA,
                                       POVERTY STRATUM on input record
                                       */
levels=vector(7,2,2,2);             /* levels for each subgroup variable */
optionvec=vector(0,0,0,0);          /* Option parms: Include, Display,
                                       Nomarg and Mergehi; */
                                    /* = "Yes", 0 = "No". */
                                    /* Table definitions using positions of
                                       variables in SUBGROUP vector * ;
tab1 = vector(2,4);                 /* RACE by POVERTY STRATUM */
tab2=vector(1,4);                   /* AGE by POVERTY STRATUM */
```

```
13 tab3=vector(1,3,4);                    /*AGE by SMSA by POVERTY
                                            STRATUM */
14 tables=vector(tab1,tab2,               /*BAT combining three tables for
                                            processing */
                  tab3);
15 records=ficsfile(filename,             /* input records with associated nest and
                                            weight variables */
                 nest var, wgtvar);
16 effvectors = effects(records,          /*create (0,1) indicator variables for
                                            subgroup levels or effects vectors */
               subgroups, levels,
               optionvec);
17 tabvec = crossp(effvectors,            /*form cross products of effects vectors
                                            */
                tables);
18 w=select(records, wgtvar);             /* input weights */
19 nsum0=sigma(tabvec,0);                 /*compute sample counts */
20 wsum0=sigma(w*tabvec,0);               /*compute weighted counts */
21 rowper = margop(wsum0,wsum0,
ROW, "/");                                /*compute row percentages */
22 tcomplement=margop(tabvec,
tabvec, Row, "-");
23 z=w*(tabvec*(1-rowper)+
tcomplement*rowper)/wsum0;                /*Taylor series limearization individual-
                                          level */
24 zpsu=sigma(z, psulev);                 /* PSU sum of Taylor series deviations */
25 zstr=sigma(zpsu, strlev);              /* Stratum sum of Taylor series deviat-
                                          ions */
26 z2str=sigma(zpsu*zpsu, strlev);        /*Stratum sum of squared deviations */
27 npsu = sigma(constant(zpsu,1),
strlev);                                  /*number of PSUs sampled within s
                                          stratum */
28 varwr = sigma((npsu*z2str-zstr*
zstr)/(npsu-1), 0);                       /*with replacement variance of row per-
                                          centage */
29 serow=sqrt(varwr);                     /*standard error of row % */
30 print(vector(rowper, serow),
vector("Row %", "S.E. of row %"));        /*prints row % and standard error of row
                                          % for the 3 tables defined in lines 11-13
                                          */
```

CROSSTAB: Statistical Formulas for SUDAAN Program Example

(1) h = stratum, $h = 1, 2, \ldots, H$
$i = PSU;\ i = 1, 2, \ldots, n_h$
j = sample unit; $j = 1, 2, \ldots, n_{hi}$
w_{hij} = weight for individual (hij)
r = row of table; $r = 1, 2, \ldots, R$
c = column of table, $c = 1, 2, \ldots, C$
p_{rc} = row percent for the rc cell

(23) $$z_{hij}(rc) = \begin{cases} 0 & \text{if sample unit } hij \text{ is not in row } r \\ w_{hij}{*}(-p_{rc'}) & \text{if sample unit } hij \text{ is in row } r \text{ and column } c' \neq c \\ w_{hij}{*}(1 - p_{rc}) & \text{if sample unit } hij \text{ is in cell } rc \end{cases}$$

(24) $$z_{hi} = \sum_j z_{hij}$$

(25) $$\sum_i z_{hi}$$

(26) $$\sum_i z_{hi}^2$$

(27) $$n_h$$

(28) $$\text{Var} = \sum_{h=1}^{H} \left[n_h \sum_i z_{hi}^2 - \left(\sum_i z_{hi} \right)^2 \right] \Big/ (n_h - 1)$$

(29) $$\text{Serow} = \sqrt{\text{Var}}$$

14

An Application of Bayesian Statistics Using Sampling/Importance Resampling for a Deceptively Simple Problem in Quality Control

T. E. RAGHUNATHAN University of Washington, Seattle, Washington

DONALD B. RUBIN Harvard University, Cambridge, Massachusetts

Bayesian procedures as implemented by simulation techniques have the potential to address complicated real world problems in their full complexity by allowing appropriately tuned models to be fit to the specific data set being studied. This important idea will be illustrated using a small but computationally complex example from quality control that involves sensitivity of inference to distributional shape. The SIR algorithm (Rubin, 1987) plays a central role in our solution.

1. SOME NOVEL USES OF BAYESIAN PROCEDURES

Many consumers of statistical techniques seem to view Bayesian statistical methods as procedures inherently hopelessly subjective—and thus nonscientific—and consequently to be avoided. In contrast, we believe that Bayesian procedures, whose hallmark is the treatment of unknown quantities as random variables and thus might be more descriptively called "full probability modeling" procedures, are the most direct path to sensible scientific inferences, especially in complicated problems. Theory and experience support the contention that well-chosen Bayesian techniques have excellent or even optimal repeated sampling properties, and so are no more subjective than the standard frequentist's choice among competing significance tests or competing robust estimators, and moreover provide more scientifically relevant answers. Such optimal sampling properties of Bayesian techniques have been established from a theoretical perspective primarily in the decision theory literature, including Berger (1986), Ferguson (1973), Maritz (1970), and Woodroofe (1976). More applied references include Dempster and Raghunathan (1987), Efron and Morris (1973), Fay and Herriot (1979), Rubin (1980, 1981), and Rubin and Schenker (1987).

Here we present an illustrative application of Bayesian statistics to a common problem that arises in quality control involving inference about the proportion of devices defective. A crucial feature of this illustration is the investigation of the sensitivity of inferences to distributional assumptions not adaquately addressed by the data set at hand. Bayesian procedures implemented by simulation techniques provide conceptually direct and computationally straightforward answers to problems such as this for which traditional answers are conceptually convoluted and computationally exhausting. A particularly novel feature of our analysis is the extensive use of the SIR algorithm (Rubin, 1987) to implement sensitivity analyses.

Although we only illustrate these sensitivity analyses in a simple quality control context, the general techniques are broadly applicable to problems of data quality control. For example, essentially all data bases suffer from problems such as missing values and systematic and random errors, where there is both uncertainty about the precise processes that created the problems and inadequate data to address the modelling of these processes. In such cases, it is highly desirable to investigate the sensitivity of inferences to different plausible models of the underlying processes, and the techniques illustrated here can be effective at implementing such sensitivity analyses.

2 SENSITIVITY TO DISTRIBUTIONAL SHAPE

Hahn and Meeker (1990, example 13.2) posit the following question: Given the surface temperatures of six randomly chosen solid-state devices—170.5, 172.5, 169.5, 174, 176, 168—estimate the proportion of devices in the population that are satisfactory in the sense that their surface temperatures will be less than 180°F. Since not one of the six devices has a surface temperature exceeding 180°F, inferences will be somewhat sensitive to assumptions about distributional shape—especially the right tail, and so an important component of a sensible statistical analysis includes an exploration of this sensitivity.

The standard model-based approach to such data posits, not surprisingly, the normal distribution [see, for example, Hahn and Nelson (1973) and Odeh and Owen (1980)], but we also consider the Cauchy distribution in order to have a distribution with much fatter tails. Also, we consider truncated versions of both the normal and the Cauchy, since it may be reasonable to assume devices with surface temperatures outside particular bounds cannot be produced: "wide" or unrestricted bounds allow values in $(-\infty, \infty)$, "medium" bounds allow values in (80, 250), whereas "narrow" bounds only allow values in (160, 190). We wish to draw inferences for the proportion of devices with surface temperatures less than 180°F under all six distributional conditions (Cauchy/normal × wide/medium/narrow). Quite clearly, the six observed values have little evidence to help choose among these six possibilities, although other related data sets may have such evidence.

2.1 Bayesian Inference Under the Unrestricted Normal Model

Bayesian inferences are very straightforward for the unrestricted normal. Let t = temperature; then we assume that given parameters $\theta = (\mu, \sigma)$

$$t|\theta \sim N(\mu, \sigma^2).$$

This is the standard normal specification for both frequentists and Bayesians. The Bayesian, however, also posits a prior distribution for θ, with density $p(\theta)$. For convenience we will take, as prior distribution for θ,

$$p(\theta) \propto \sigma^{-1}.$$

See, for example, Box and Tiao (1973) for extensive discussion of such prior distributions. We are primarily concerned with sensitivity to the distributional shape of t given θ, which is far more important in this example than sensitivity to the prior distribution on θ. It would be easy, however, to extend our sensitivity analyses to various forms of $p(\theta)$, for example, with bounds on the possible values of θ.

Given the six observed temperatures $\mathbf{t} = (t_1, \ldots, t_6)$ we have a posteriori:

$$\sigma^2|\mathbf{t} \sim SS/\chi_5^2, \tag{2.1.1}$$

Table 1. Percentiles of Simulated Posterior Distributions of $\pi(\theta) = \Pr(t < 180|\theta)$ = proportion satisfactory, in %

Percentile	Normal	T, Normal (80,250)	T, Normal (160,190)	Cauchy	T, Cauchy (80,250)	T, Cauchy (160,190)
0.005	22[a]	22	23	28	30	33
0.010	31	35	34	36	37	42
0.025	51	50	54	50	52	58
0.050	70	64	67	64	65	69
0.100	84	82	83	75	76	80
0.250	96	95	96	84	85	89
0.500	100[b]	100	100	88	90	93
0.750	100	100	100	91	92	95
0.900	100	100	100	93	94	96
0.950	100	100	100	94	94	97
0.975	100	100	100	94	95	97
0.990	100	100	100	95	95	97
0.995	100	100	100	95	95	98

[a] E.g., Pr (percentage satisfactory <22%) = 0.005 = Pr (percentage satisfactory at least 78%) = 0.995

[b] All 100's mean >99 but <100.

where

$$SS = \sum (t_i - \bar{t})^2,$$
$$\bar{t} = \sum t_i/6,$$
$$\chi_5^2 = \text{a chi-squared random variable on 5 degrees of freedom,}$$

and

$$\mu|\sigma^2, \mathbf{t} \sim N(\bar{t}, \sigma^2/6). \qquad (2.1.2)$$

Now the proportion of devices with surface temperatures less than 180°F is a function of θ, say $\pi(\theta)$, which for the unrestricted normal is

$$\pi(\theta) = \Phi\left(\frac{180 - \mu}{\sigma}\right), \qquad (2.1.3)$$

where $\Phi(\ )$ is the normal cumulative distribution function. Although analytic expressions for the posterior distribution of $\pi(\theta)$ can be presented using (2.1.1) and (2.1.2), it is just as instructive, and more relevant to inference under the other distributions we will be considering, to describe how to simulate the posterior distribution of $\pi(\theta)$.

First draw a value of θ, say θ_* following (2.1.1) and (2.1.2), by first drawing σ_* using a χ_5^2 random variable (i.e., $\sigma_*^2 = SS/\chi_5^2$), and second drawing μ using the drawn value of σ_* and a standard normal deviate (i.e., $\mu_* = \bar{t} + \sigma_* N(0, 1)$), and then calculate $\pi_* = \pi(\theta_*)$ using (2.1.3). Repeated draws of θ yield a simulated posterior distribution for $\pi(\theta)$; percentiles of this posterior distribution based on 10,000 draws are given in the second column of Table 1. This can be done very quickly even on a modest personal computer.

2.2 BAYESIAN INFERENCE WITH NONSTANDARD DISTRIBUTIONS

Inference, Bayesian or otherwise, for $\pi(\theta)$ under the Cauchy or truncated normal distribution is analytically unpleasant. Simulation techniques, however, can make Bayesian inference with such distributions quite straightforward, especially when we use the SIR algorithm to draw a sample from the appropriate posterior distributions of $\pi(\theta)$.

Before describing this applicatiom of SIR or defining SIR itself, we note that $\pi(\theta)$ is an easily written function of θ, so that the real issue in simulating $\pi(\theta)$ is simulating θ since given a drawn value of θ, θ_*, the corresponding drawn value of π is $\pi(\theta_*)$. For the truncated normal with $L < t < U$, that is, for the density

$$p(t|\theta) = \begin{cases} \sigma^{-1}\phi\left(\dfrac{t-\mu}{\sigma}\right)\Big/\left[\Phi\left(\dfrac{U-\mu}{\sigma}\right) - \Phi\left(\dfrac{L-\mu}{\sigma}\right)\right] & \text{if } L < t < U \\ 0 & \text{if } t \leqslant L \text{ or } t \geqslant U, \end{cases}$$

where $\phi(z) = 1/\sqrt{2\pi}\exp(-z^2/2)$ and $\Phi(x) = \int_{-\infty}^{x} \phi(z)\,dz$, we have

$$\pi(\theta) = \frac{\Phi\left(\dfrac{180-\mu}{\sigma}\right) - \Phi\left(\dfrac{L-\mu}{\sigma}\right)}{\Phi\left(\dfrac{U-\mu}{\sigma}\right) - \Phi\left(\dfrac{L-\mu}{\sigma}\right)}, \tag{2.2.1}$$

which reduces to (2.1.3) when $L = -\infty$ and $U = +\infty$.

For the Cauchy distribution with location μ and scale σ, that is, for the density $\sigma^{-1}C([t-\mu]/\sigma)$ where $C(z) = [\pi(1+z^2)]^{-1}$,

$$\pi(\theta) = \frac{1}{2} + \frac{1}{\pi}\tan^{-1}\left(\frac{180-\mu}{\sigma}\right), \tag{2.2.2}$$

and for the truncated Cauchy

$$\pi(\theta) = \frac{\tan^{-1}\left(\dfrac{180-\mu}{\sigma}\right) - \tan^{-1}\left(\dfrac{L-\mu}{\sigma}\right)}{\tan^{-1}\left(\dfrac{U-\mu}{\sigma}\right) - \tan^{-1}\left(\dfrac{L-\mu}{\sigma}\right)}, \tag{2.2.3}$$

which reduces to (2.2.2) when $L = -\infty$ and $U = +\infty$. Thus, if we had draws of θ from its posterior distributions corresponding to truncated normal and truncated Cauchy specifications, simulated posterior distributions of $\pi(\theta)$ would be immediate using (2.2.1) and (2.2.3).

The posterior distribution of $\theta = (\mu, \sigma)$ under the truncated normal specification with $p(\theta) \propto \sigma^{-1}$ is

$$p_{\text{TN}}(\mu, \sigma|\mathbf{t}) \propto \prod_{i=1}^{6} \phi\left(\frac{t_i-\mu}{\sigma}\right)\Big/\left\{\sigma\left[\Phi\left(\frac{U-\mu}{\sigma}\right) - \Phi\left(\frac{L-\mu}{\sigma}\right)\right]\right\}^6 \sigma^{-1}\,d\mu\,d\sigma \tag{2.2.4}$$

which is not an easy distribution from which to draw (μ, σ) except when $L = -\infty$ and $U = +\infty$.

Similarly, the posterior distribution of θ under the truncated Cauchy specification is

$$p_{\mathrm{TC}}(\mu, \sigma|\mathbf{t}) \propto \prod_{i=1}^{6} C\left(\frac{t_i-\mu}{\sigma}\right) \Big/ \left\{\sigma\left[\tan^{-1}\left(\frac{U-\mu}{\sigma}\right) - \tan^{-1}\left(\frac{L-\mu}{\sigma}\right)\right]\right\}^6 \sigma^{-1}\, d\mu\, d\sigma \qquad (2.2.5)$$

which is not an easy distribution from which to draw (μ, σ) even when $L=-\infty$ and $U=+\infty$.

2.3 Drawing Values of θ Using SIR

Our plan is to obtain draws from the posterior distributions (2.2.4) and (2.2.5) using the following facts:

1. Drawing (μ, σ) under the unrestricted normal is easy (and done already in Section 2.1).
2. For any drawn value of (μ, σ), the posterior density of (μ, σ), up to a constant, is easy to calculate using (2.2.4) or (2.2.5).

Specifically our plan is to follow two steps that, in this case, define the SIR (sampling/importance resampling algorithm) proposed in Rubin (1987):

1. For each of the 10,000 values of θ drawn under the unrestricted normal model, calculate the importance ratio (up to a multiplicative constant) for each of the five other models:

$$\text{Importance ratio }(\theta_*) = \frac{\text{posterior density of } \theta_* \text{ under desired model}}{\text{posterior density of } \theta_* \text{ under unrestricted normal model}}. \qquad (2.3.1)$$

For example, for the truncated Cauchy specification

$$\text{Importance ratio }(\theta) \propto \frac{\Pi_{i=1}^{6} C\left(\frac{t_i-\mu}{\sigma}\right) \Big/ \left[\tan^{-1}\left(\frac{U-\mu}{\sigma}\right) - \tan^{-1}\left(\frac{L-\mu}{\sigma}\right)\right]^6}{\Pi_{i=1}^{6} \phi\left(\frac{t_i-\mu}{\sigma}\right)}. \qquad (2.3.2)$$

2. For each model, draw 2000 values of θ_*, or equivalently $\pi(\theta_*)$, from the 10,000 values with probability (approximately) proportional to the model's importance ratios (Cochran, 1977, chapter 9A). For example, we list the 10,000 values of θ_* and construct the associated cumulative sum of the importance ratios. To draw the first value, generate a uniform random number u and choose the θ_* that corresponds to the cumulative sum closest, but less than or equal, to ku, where $k = \Sigma_{10,000}$ importance ratios. To generate the second θ_*, we eliminate the drawn value from consideration and subtract the importance ratio that corresponds to the selected θ_* in the previous draw from k and proceed as in the previous step to draw from the remaining 9999 θ_* values. Thus, the 2000 values of θ_*, and hence the $\pi(\theta_*)$, are drawn without replacement from the 10,000 values with probability approximately proportional to the size of the importance ratios.

The 2000 drawn values of $\pi(\theta_*)$ are approximate draws from the correct posterior distribution. The justification is given in Section 3, but we examine results before presenting this justification.

2.4 Results of SIR Algorithm

The last five columns of Table 1 summarize the results of 2000 draws of π = proportion of devices with surface temperature $<180°F$ from each of the five models. The same table values would have been obtained using the importance ratios as weights to be applied to the 10,000 values, but SIR has several virtues that arise from having a simple random sample from the posterior distribution.

There is apparently some sensitivity to the range of models considered. For example, from the fifth percentile we see that under the most pessimistic model the probability is .95 that at least 64 percent of the devices have surface temperatures less than 180°F and thus are satisfactory; under the normal specification the probability is .95 that at least 70 percent of the devices are satisfactory. With .50 probability, there is more sensitivity. For example, across all the models it is at least an even bet that at least 88 percent of the devices are satisfactory, whereas if one accepted normality, it is an even bet that over 99.5 percent of the devices are satisfactory.

Given that the six observed data points have very limited information about distributional shape, in the absence of other information it would seem wisest to be conservative and make decisions with the understanding that it is quite possible, even a coin toss, that fewer than 88 percent of the devices are satisfactory.

Notice how easily these Bayesian inferences were obtained using simulation implemented by the SIR algorithm: Draw many (10,000) values under the standard easy model, and then resample some (2000) using the easily calculated importance ratios. We next turn to a problem of constructing tolerance bounds.

2.5 Extensions to Tolerance Intervals

The parameter $\pi(\theta)$, or in a more explicit notation, $\pi_{180}(\theta)$, is the proportion of satisfactory devices in a lot, where satisfactory in the context of Section 2.4 means $t \leqslant 180$. In the statement of the practical problem by Hahn and Meeker (1990), it was hoped, from the data, we could conclude with high confidence that at least 99 percent of the devices would be satisfactory. Since the results of Hahn and Meeker's analyses (and our analyses, too) suggest that it is quite possible that far less than 99 percent of the devices are satisfactory in this sense of having $t < 180$, attention is redirected at finding a definition of satisfactory, that is, a temperature $t(\theta)$, such that 99 percent of the devices are satisfactory; in a more explicit notation, we seek $t_{99}(\theta)$ such that

$$\int_{-\infty}^{t_{99}(\theta)} f(t|\theta)\,dt = 99\% \,. \tag{2.5.1}$$

Interval estimates for $t(\theta) = t_{99}(\theta)$ are traditionally called tolerance intervals. From our Bayesian perspective, the objective is to simulate the full posterior distribution of $t(\theta)$. Specifically, for each of the six models and each of the 2000 values of θ drawn to construct Table 1, instead of calculating $\pi(\theta)$ and ordering the 2000 values as was done for Table 1, we now calculate $t_{99}(\theta)$ and order them. For all models, $t_{99}(\theta)$ is a straightforward function of θ; for instance, for the truncated Cauchy

$$t_{99}(\theta) = \mu + \sigma \tan\left[0.99 \tan^{-1}\left(\frac{U-\mu}{\sigma}\right) + 0.01 \tan^{-1}\left(\frac{L-\mu}{\sigma}\right)\right].$$

The results of the computations are displayed in Table 2, which gives the percentiles of the posterior distribution of $t_{99}(\theta)$ under various models. For example, with probability 90 percent at least 99 percent of the devices have temperature not exceeding 186.3°F under the unrestricted normal model, whereas for the truncated (80, 250) Cauchy model, the corresponding temperature is 226.8°F. Clearly, there is substantial sensitivity to the range of models considered. The results are more sensitive to truncation under the Cauchy model than the normal models. This is expected because the tail probability outside (L, U) interval is larger under the Cauchy

Table 2. Percentiles of the Posterior Distributions of $t_{99}(\theta)$ Under Various Models

Percentile	Normal	T, Normal (80,250)	T, Normal (160,190)	Cauchy	T, Cauchy (80,250)	T, Cauchy (160,190)
0.005	170.7	170.6	170.3	281.1	223.0	189.5
0.010	171.7	171.7	170.4	281.2	223.0	189.5
0.025	172.9	172.8	171.8	281.5	223.1	189.5
0.050	173.7	173.7	176.7	282.0	223.2	189.5
0.100	174.8	174.8	173.8	283.0	223.4	189.5
0.250	176.6	176.6	176.2	285.8	224.1	189.6
0.500	179.0	179.0	179.0	290.6	225.1	189.6
0.750	182.2	182.2	182.7	295.4	226.2	189.7
0.900	186.3	186.3	188.0	298.2	226.8	189.8
0.950	189.5	189.5	189.7	299.2	227.0	189.8
0.975	193.5	193.5	189.9	299.6	227.1	189.8
0.990	199.0	199.0	189.9	299.9	227.2	189.8
0.995	202.5	202.5	189.9	300.0	227.2	189.8

model than under the normal model. Hahn and Meeker (1990) only consider tolerance intervals under the unrestricted normal and obtain as the ninetieth percentile 185°F, very close to our 186°F.

2.6 Posterior Predictive Distributions of Temperature for a New Device

A final analysis for this data set involves the consideration of the predictive distribution of the temperature for a new randomly chosen device, that is,

$$\Pr(t|\mathbf{t}) = \int \Pr(t|\theta) \Pr(\theta|\mathbf{t})\, d\theta.$$

To simulate the posterior predictive distribution of temperature for a new device t, for each value of θ_* under each model, we generate repeated values of t from the appropriate distribution of temperature under that model, $\Pr(t|\theta = \theta_*)$. For example, under the unrestricted normal, we let

$$t_* = \mu_* + \sigma_* z, \tag{2.6.1}$$

where z is a standard normal deviate ($z \sim N(0, 1)$); under the Cauchy model restricted to (L, U), we let

$$t_* = \mu_* + \sigma_*(z_1/z_2) \tag{2.6.2}$$

if t_* is between L and U, and redraw (z_1, z_2) if it is not, where z_1 and z_2 are independent standard normal deviates.

Figures 1 and 2 give stem-and-leaf displays of these distributions for the normal and Cauchy models respectively based on 200 draws. These indicate more sensitivity to truncation than to distributional assumptions, although the posterior predictive distributions have fatter tails under the Cauchy models than under the normal models. Medians of the posterior predictive distributions are quite similar under the six different models. Comparable displays of sensitivity of inference under standard frequentist approaches would be extremely difficult to produce.

3. JUSTIFICATION FOR THE SIR ALGORITHM

Suppose we wish to draw m values from the density, $p(u)$, of the vector random variable u (e.g., the posterior density of a parameter θ given data x). In many instances it is not possible to calculate, let alone draw from, this density. Such situations are abundant in the incomplete data literature.

Stem	Unrestricted	Truncated (80,250)	Truncated(160,190)
		5	
157	8	8	
158	12		
159	1	4	
160	7	2358	189
161		6	368
162	1	689	26667
163	49		45
164	134569	4689	114589
165	01123466	267	023456778
166	057	0257778	0035577
167	0111223355667888999	001233445	0014666668899
168	0344456699	0133444567888889	0234444555689
169	1233344445667788	0111224447889	02556668
170	000001133355566667888	000000111224455556667788888899	0001333466666777889999
171	00011234446788999	00111112235566999	00011333444456678999
172	00223444556777899	00012333444555566788	0033444555677789999
173	0122345666888999	0111334556889	012234578999
174	01123345566889	1222455567799	012233334556677778889
175	01112349	0013344558	00144468
176	1113344778	113344699	1112344788889
177	012339	3369	0224999
178	11111458	01445678	14
179	359	4	5
180	017	28	
181	9	08	2
182		125	2
183	7	3	1
184	47	0	0
185		1	7
186	49		
187		5	8
188		3	3
189			
190			
191			
192			
193	9		

Unrestricted Truncated (80,250) Truncated(160,190)

Figure 1. Stem-and-leaf displays of the simulated predictive distributions of temperature under various normal models.

Stem	Unrestricted	Truncated(80,250)	Truncated(160,190)
150	5	5	
151			
152	8	8	
153			
154			
155	8	8	
156	7	7	
157	67	67	
158	12	12	
159	14	14	
160	08	8	358
161	27	27	278
162	2236679	22223679	22223679
163	26679	26679	2456679
164	16789	16788	16789
165	0012466668	0012466668	0012466668
166	0667	0667	066778
167	0001134555569	00001145555669	00014555566899
168	022333455677899	02233345555555677899	02222333455556778999
169	0011244556789	00112444455567789	00112445556778?
170	00366688999	01266688999	026688999
171	00001146	00001146	0000011469
172	0179	00157	001467
173	0333355778889	01333557889	03333557889
174	0001224778899999	0001222457789999	0012224577889999
175	12456677789	12456889	12456777899
176	124457889	1244588	122444588
177	035689	035689	03689
178	111144578	111144578	11114578
179	14799	14799	147899
180	3458	358	34578
181	269	2699	269
182	125	125	125
183	17	1	17
184	03	03	03
185	7	7	7
186			
187	58	58	58
188			
189			
190	559	559	
191			
192			
193	9	9	
194	8	8	
200	5		

Figure 2. Stem-and-leaf displays of the simulated predictive distributions of temperature under various Cauchy models.

Let v be an auxiliary vector random variable such that $p(u, v)$, the joint density of (u, v), is easily evaluated up to a multiplicative constant free of (u, v). Such "auxiliary" or "nuisance" variables can be found in many situations. For example, with incomplete data, the joint posterior density of parameter θ and missing data, X_{mis}, given the observed data, X_{obs}, is easily evaluated, since it is proportional to the density of the complete data $(X_{\text{mis}}, X_{\text{obs}})$ times the prior density of θ.

Furthermore, let $h(u, v)$ be an approximation to $p(u, v)$ such that (a) it is easy to draw (u, v) from it, and (b) it can be easily evaluated up to a multiplicative constant.

The following steps provide m approximate draws from the density $p(u)$.

Step I: Draw M values from $h(u, v)$, say (u_*, v_*), $M > m$.
Step II: Attach importance ratios to the drawn values of (u, v). That is,

$$w_* \propto \frac{p(u_*, v_*)}{h(u_*, v_*)}.$$

Step III: Resample m values of u_* from the M values of u_* with probability proportional to w_*. Equivalently, resample m values of (u_*, v_*) with probability proportional to w_* and discard the values of v_*.

Note that as $M/m \to \infty$, the draws of (u_*, v_*) obtained at step III are from the joint density $p(u, v)$:

$$\Pr(u_*, v_*) = \frac{h(u_*, v_*)w_*}{\int h(u_*, v_*)w_* d(u_*, v_*)} = p(u_*, v_*).$$

Therefore, by ignoring or discarding the values of v_* we obtain m draws from the posterior density $p(u)$, at least when $M/m \to \infty$.

Also, note that the example discussed in Section 2.4 uses the SIR algorithm with $u = \pi(\theta)$ and $v = \theta$, which implies a degenerate joint distribution for (u, v):

$$\Pr(u|v) = \begin{cases} 1 & \text{if } u = \pi(v) \\ 0 & \text{otherwise}. \end{cases}$$

In Section 2.5, $\Pr(u|v)$ is also degenerate with $\mu = t_{99}(\theta)$ and $v = \theta$; and in Section 2.5, $\Pr(u|v)$ is the conditional distribution of the temperature of a new device given v. For all subsections of Section 2, we used the normal/inverted chi-squared density for $h(v)$, common for all models, and degenerate distributions of u given $v, h(u|v)$ for $\pi(\theta)$ and $t_{99}(\theta)$, but the appropriate distribution, $h(u|v)$, for the temperature t.

Issues needing further study include: What makes a good approximating distribution $h(u, v)$ (e.g., it should have fatter tails than $p(u, v)$); what's an appropriate ratio for M/m; are adaptive versions of choosing M/m better (e.g., based on the variance of the first m log importance ratios); what are the relative benefits of sampling with and without replacement; and what about iterative versions of SIR? Some of these are very briefly discussed in Rubin (1988).

REFERENCES

Berger, J. O. (1986). *Statistical Decision Theory and Bayesian Methods*, Springer-Verlag, New York.

Box, G. E. P., and Tiao, G. C. (1973). *Bayesian Inference in Statistical Analysis*, Addison-Wesley, Reading, MA.

Cochran, W. G. (1977). *Sampling Techniques*, Wiley, New York.

Dempster, A. P., and Raghunathan, T. E. (1987). Using a covariate for small area estimation: A common sense Bayesian approach, *Proceedings of Small Area Statistics, An International Symposium*, eds.: Platex, R. Rao, J. N. K., Särndal, C. E. and Singh, M. P.

Efron, B., and Morris, C. (1973). Stein's estimation rule and its competitors—An empirical Bayes approach." *Journal of the American Statistical Association.*

Fay, R. E., and Herriot, R. A. (1979). Estimates of income for small places: An application of James-Stein procedures to census data, *The Journal of American Statistical Association 74*, 269–277.

Ferguson, T. S. (1973). *Mathematical Statistics; A Decision-Theoretic Approach*, Academic Press, New York.

Hahn, G. J., and Nelson, W. (1973). A survey of prediction intervals and their applications, *Journal of Quality Technology 5*, 178–188.

Hahn, G. J., and Meeker, W. Q. (1990). *Statistical Intervals: A Guide for Practitioners*, Wiley, New York.

Maritz, J. S. (1970). *Empirical Bayes Methods*, Methuen, London.

Odeh, R. E., and Owen, D. B. (1980). *Tables for Normal Tolerance Limits, Sampling Plans and Screening*, Marcel Dekker, Inc., New York.

Rubin, D. B. (1980). Using empirical Bayes techniques in the Law School Validity Studies, *The Journal of the American Statistical Association 75*, 801–827.

Rubin, D. B. (1981). Estimation in parallel randomized experiments, *Journal of Educational Statistics 6*, 377–400.

Rubin, D. B. (1982). The SIR algorithm—A discussion of Tanner and

Wong's "The calculation of posterior distributions by data augmentation." *Journal of the American Statistical Association 82*, 543–546.

Rubin, D. B. (1988). Using the SIR algorithm to simulate posterior distributions. *Bayesian Statistics, Vol. 3*, Bernardo, DeGroot, Lindley and Smith (eds). Oxford University Press, 395–402.

Rubin, D. B., and Schenker, N. (1987). Interval estimation from multiply-imputed data: A case study using agriculture industry codes, *Journal of Official Statistics*.

Woodroofe, M. (1976). Frequentist properties of Bayesian sequential tests, *Biometrika 63*, 101–110.

15
Quality Issues and Evidence in Statistical File Merging

RICHARD S. BARR Southern Methodist University, Dallas, Texas

J. SCOTT TURNER College of Business Administration, Oklahoma State University, Stillwater, Oklahoma

1. BACKGROUND AND OVERVIEW

The concept of microanalytic simulation models was developed by Guy Orcutt in the mid-1950s (Greenberger et. al, 1976). Today, these models abound in governmental agencies and research organizations and are used widely for policy analysis and projection of program needs. Examples include the various versions of the Transfer Income Model (TRIM); the dynamic, demographic simulation model DYNASIM; and the tax-policy simulations at the U.S. Department of the Treasury, Brookings Institution, and Statistics, Canada.

At the heart of these models are sample survey files, or microdata. These files consist of data records for a representative set of decision units (individuals, households, taxpayers, firms, etc.) that are processed by the simulator individually with data collected to identify aggregates, distributions, and interactions. By working at the record level, this modeling technique is very flexible and can accommodate as much detail as desired.

Microdata matching is useful when the data required for a model are located in two or more microdata sets, as is often the case with governmental tax- or welfare-system models. Means for performing such matchings are the subject of this chapter. At present, all known applications for microdata matching are in the public sector, primarily at the national level. However, these same methodologies could be used in business applications such as the construction of marketing research databases from a set of separately compiled samples.

1.1 Microdata Files

While the recording unit may vary, microdata files usually represent the national population or a major subset such as taxpayers or Social Security system participants. Since various sampling schemes are used in collecting the data, each record includes a weight indicating the number of population units it represents. These weights often differ among records in a given file.

Microdata files are created as byproducts of ongoing governmental programs, from legislative mandate, or as special commissioned studies. For example, both the IRS's Statistics of Income (SOI) and Social Security Earnings (SSA) files are drawn from data collected in the process of program implementation and control. The U.S. Constitution mandates the taking of a decennial census, subsets of which are used as microdata, and the Current Population Survey (CPS) is performed monthly to determine the unemployment rate, as required by law. The Survey of Income and Education (SIE) was a special study, as are numerous university-based surveys.

For the model designer and user, there are several pertinent characteristics of microdata files. First, they are expensive to create, on the order of $10 million each. Hence, their construction is not a trivial undertaking. Second, several versions are often created through editing procedures to "correct" the data for underreporting, sample bias, etc. Third, a variety of sampling designs may be used, including stratified, clustered, and simple randomized, in order to combine information richness with brevity. Fourth, the end product of these sometimes elaborate machinations is a multiattribute representation of the underlying population, including all interactions and distributions of the reported data items. The distributional and interaction details are especially important for microanalytic models, since they operate at the record level and base their computations on combinations of item values. Finally, by virtue of taking a middle ground between a census and population aggregates, these files are efficient from both a computational and information-content standpoint.

1.2 Limitations of Individual Samples

As illustrated by files such as the SOI and CPS, microdata are often collected primarily for the construction of aggregates or for program implementation, analysis, and control. Their use as general research databases or in microanalytic simulation models is of secondary concern in the sample survey designs, an aspect that creates problems for these applications.

As models are built and policy proposals are analyzed, data are often required that (1) are not part of the current program, study, or system, as when new tax deductions are considered; or (2) are of superior quality, since sample survey items are deemed to be unreliable, such as business income on the CPS.

The model user has four choices available: (1) commission a new study, at great expense and investment of time; (2) ignore the variables in question, and jeopardize the validity of the model's results; (3) impute the missing or unreliable items into an existing file, using methods that often ignore the distributional and interaction characteristics of the variables in question; or (4) merge a pair of microdata files to combine the information from two surveys. This last option, file merging, is currently in widespread use and is investigated in this chapter.

1.3 Microdata File Merging

The basic idea behind file merging, or matching, is to combine one file A with another file B to form a composite file C with all data items from the two original files. This is accomplished by selecting pairs of records to match based on data items that are common to both files. The schemes for performing the matching process fall into two general categories: exact and statistical matching.

Exact matching uses unique-valued common items to mate records for the same individual in both files. By using a unique identifier, such as a social-security number, the matching process is theoretically a simple sort and merge operation. Problems with this approach include insignificant overlapping of samples, causing few records to be matched; absence of or error in the "unique" identifiers; confidentiality restrictions that preclude legal linking of records; and the expense of handling a large number of exceptions.

Statistical merging (also called synthetic, stochastic, or attribute matching or merging) mates *similar* records using several common items with nonunique values. By matching like records, file C contains records that may be composites of two different persons, but whose attributes are similar enough for research purposes. There are a variety of statistical merging schemes in use today, as discussed below.

In choosing a methodology, exact matching is obviously preferable. But where such a match is not possible, statistical merging is often employed.

1.4 Statistical Merging

A pictorial description of statistical merging is presented in Figure 1. In this drawing, a_i represents the weight of the ith record in file A and b_j represents the weight of the jth record in file B. The merged file, C, contains composite records formed by matching a record in file A with a record in file B and assigning a merge record weight of w_{ij}. An interrecord dissimilarity measure, d_{ij}, or distance function, is used to choose matched record pairs. The "distance" between a pair of records is usually determined from a user-defined function that compares corresponding common items and assigns a penalty value for each item pair that differs significantly. These penalties are summed to create a measure of dissimilarity, with a zero distance meaning that all common items are identical or close enough.

There are two general categories of statistical merges: unconstrained and constrained. In an unconstrained merge, file A is designated the base file and file B the augmentation file. Each base-file record is matched with the most similar record in the augmentation file; the selected file-B record is appended to the base-file record and the base record's weight is used for w_{ij}. This is, in essence, sampling with replacement, since some augmentation file records may not be matched, while others may be used repeatedly. This is a very popular technique, as evidenced by its use by Ruggles and Ruggles (1974) of Yale and NBER, Radner (1980) of the Social Security Administration, Okner (1974) and Minarik (1980) at Brookings Institution, Statistics Canada (Colledge et al., 1979) and the Bureau of the Census.

In contrast, a constrained merge uses matching without replacement. The merging algorithm enforces constraints on the record weights in both files to ensure that each record is neither under- nor over-matched relative to the number of population units represented. Mathematically, the constrained merge model is as follows:

$$\sum_{i=1}^{m} a_i = \sum_{j=1}^{n} b_j \qquad (1)$$

$$\sum_{j=1}^{n} w_{ij} = a_i, \qquad i = 1, \ldots, m \tag{2}$$

$$\sum_{i=1}^{m} w_{ij} = b_j, \qquad j = 1, \ldots, n \tag{3}$$

$$w_{ij} \geq 0, \qquad \text{for all } i \text{ and } j \tag{4}$$

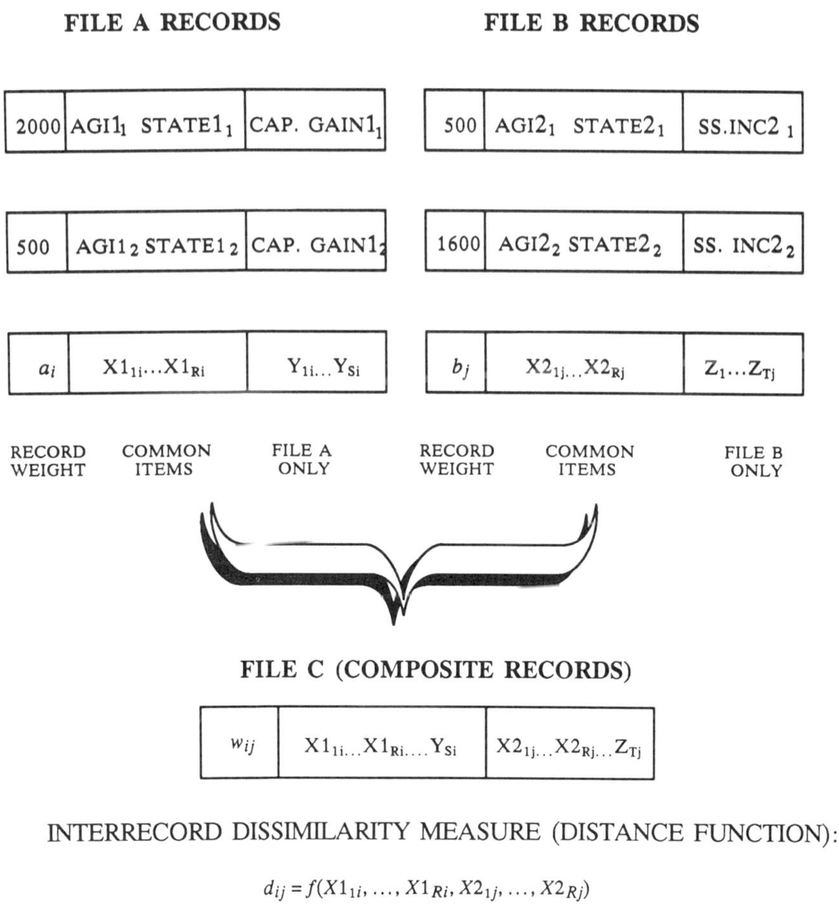

Figure 1. Statistical file merging.

Constraint (1) reflects the assumed equivalent underlying population sizes for the two files, although files A and B have m and n records, respectively. Some minor adjustments may be needed to accomplish this in practice. Again, w_{ij} is the merged record weight for matching record i in file A with record j in file B, and the records are not matched if $w_{ij} = 0$. Constraints (2) and (3) allow any record to be matched one or more times, but such that the merge file weights must sum to the original record weights. Negative weights are precluded by (4). This merging algorithm is currently used by Mathematics Policy Research.

Pictorially, the constrained merge process is depicted in Figure 2, where the leftmost set of circles, or nodes, represents the file-A records with

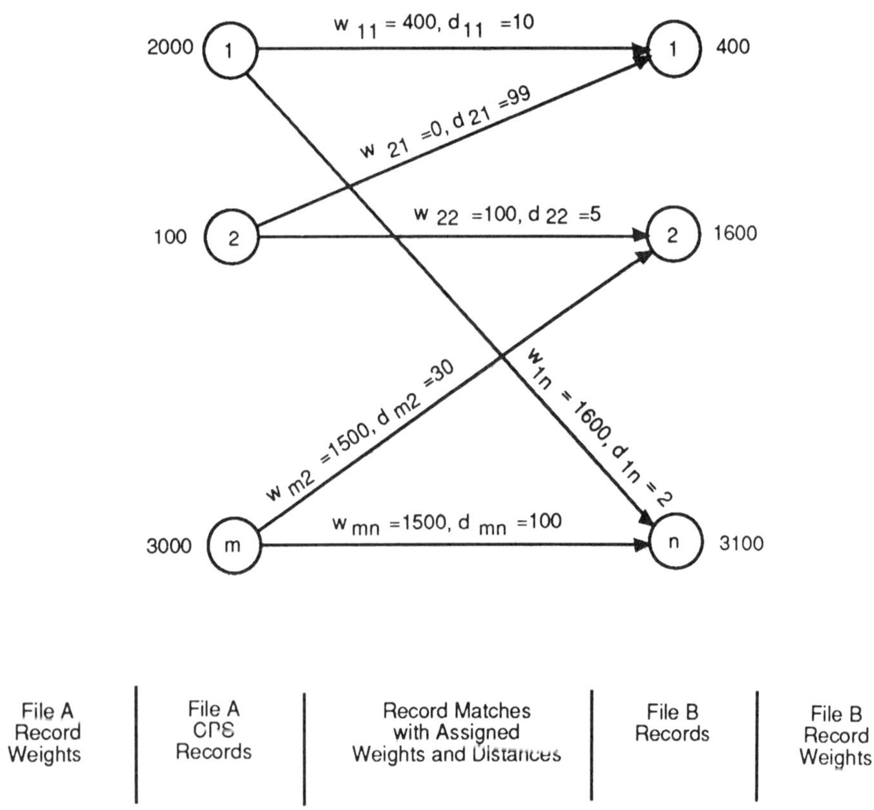

Figure 2. Constrained merge model.

their respective weights, the rightmost nodes represent the file-B records and weights, and the connecting arcs represent the possible record matches. A set of w_{ij} merge record weights is shown that meets constraints (1) through (4).

This merge technique can be further refined by requiring the procedure to

$$\text{minimize} \sum_{i=1}^{m} \sum_{j=1}^{n} d_{ij} w_{ij} \tag{5}$$

subject to (1) through (4). This model, originally proposed by Turner and Gilliam (1975) and later derived by Kadane (1975), seeks to find the best constrained match, the one with the minimum aggregate distance between matched records. This optimal constrained merge procedure requires the solution of a linear programming problem of extremely large dimensions and is currently used by the U.S. Department of the Treasury (Barr, 1982; Barr and Turner, 1978; Barr and Turner, 1980; Barr and Turner, 1981).

1.5 Statistical Aspects of Merging Techniques

Unconstrained procedures use constraint (5), subject to constraints (1), (2), and (4); thus, by dropping constraint set (3), the composite records match up well at the record level. However, file-B item statistics in the composite file are distorted by their implicit reweighting of the augmentation records through over- and under-matching. This reweighting has a strong impact on important extreme values, variances, covariances, and other distributional aspects of the file-B items, as shown below.

While the constrained procedures may not match up as well at the record level as do unconstrained procedures, their merged files contain all the information from the original files and preserve all statistical properties of the A and B data items. Further, if optimization is applied, the best overall constrained match is ensured. The appendix details, using two small example files, the effect of various merging schemes on the mathematical structure of the resultant composite file.

All these aspects influence the results of the microanalytic models and research studies that use the merge file.

1.6 Underlying Merge Rationale

When two files are merged, we assume that two files $(X1, Y)$ and $(X2, Z)$ are drawn from the same population, where $X1$ and $X2$ are the sets of common items in files A and B, respectively, and Y and Z the sets of items unique to files A and B, respectively (the alignment assumption). The objective of merging is to form a file $(X1, X2, Y, Z)$ that corresponds statistically to a sample of (X, Y, Z) taken from the same population. We do this to make inferences about (Y, Z) and $(Y, Z|X)$ relationships, since we can already make (X, X), (X, Y), and (X, Z) inferences from the two original files.

1.7 Quality Considerations

A strong theoretical justification for merging or an explanation of exactly what is being accomplished by a merge is not available in the literature. What is needed is both a measure of the "accuracy" of $(X1, X2, Y, Z)$ in replicating (X, Y, Z) and a means for making decisions such as: Are the two files mergeable? Is the composite file acceptable?

Typical reported measures of match accuracy are counts of $X1$-$X2$ item agreements, item means, and percentage agreement by common item. The notion is that a file that matches well on the X items matches well on the Y-Z relationships. Rarely reported are (1) comparisons of covariances, such as $\text{cov}(Y)$ in unconstrained matches and $\text{cov}(Y, Z)$ versus expected $\text{cov}(Y, Z)$; (2) conditional and joint frequencies for augmentation variables; and (3) other Y-Z studies. Ruggles, Ruggles, and Wolff (1977) are the only contributors in this area. Moreover, the following empirical data direct attention to potential problems measured by the above-mentioned statistical measures.

1.8 PRELIMINARY EMPIRICAL DATA

In a recent set of experiments to investigate the effect of merging techniques on resultant file quality, subsets of the 1975 SOI and 1975 SIE were chosen, based on a nine-state geographic region. There were 7144 SOI records and 6283 SIE records, representing 12.7 million tax-filing units, or approximately 15 percent of the population. The X variables used in the distance function were age, race, sex, marital status, family size, wage income, business income, property income, spouse's income, and adjusted gross income. The two files were merged three ways: unconstrained with SOI as the base file, unconstrained with SIE as the base file, and optimally constrained, all using the same distance function.

Table 1. SOI Item Distributions

Income Class ($Thousands)	Total SOI Wages ($ Millions) Original	Unconstrained	Total SOI Business Income ($ Millions) Original	Unconstrained
<0	0	0	−712	−86
1–5	12,218	11,699	857	607
5–10	22,535	21,124	836	1,194
10–15	24,745	26,639	617	1,497
15–20	10,326	23,882	677	909
20–30	21,133	24,930	808	1,402
30–50	9,597	10,141	785	964
50–100	3,371	2,741	777	1,108
100–200	1,010	136	298	608
>200	244	0	111	0
Total	115,784	121,291	5,055	8,187
Mean	$9,108	$9,542	$398	$644

Table 2. Variance-Covariance Differences

Common Variable	Median Absolute Variance-Covariance Error Relative to Original Data, %	
	Unconstrained SIE	Unconstrained SOI
Age	31.2	26.4
Family size	35.5	17.4
Wages	7.3	23.9
Business income	72.1	38.4
Farm income	97.7	88.8
Property income*	78.5	850.4
Spouse income	73.5	31.0
Adjusted gross income	9.9	24.4

*Interest + dividend + rental income.

In Table 1, the distribution of SOI wages and business income is shown for both the original file and the unconstrained merge file using the SIE as the base. Not only are the means not in agreement, but the distributions are altered, and dramatically in the case of business income. Of course, in the constrained merge, the distributions are identical to those of the originals.

To evaluate the unconstrained procedure's effect on covariance structure, the variance-covariance matrices of several common items were compared with the original matrices. The median absolute percentage differences, by item, are shown in Table 2. In one case, the median error is 7 percent, but in others these second-order statistics differ greatly. Analysis of the constrained merge verified the expectation of zero error.

1.9 Research Questions

Despite the widespread use of merging as a data-enrichment technique, there is a paucity of much-needed research in this area. Consider the following questions.

Constrained Versus Unconstrained Techniques

When does either procedure create a match file that is statistically equivalent to a valid (X, Y, Z) sample drawn from the population? A goodness-of-match criterion is needed not only to answer this question, but to compare alternative matching algorithms.

Covariance of $(Y, Z|X)$

What is the effect of omitting or including $\text{cov}(Y, Z|X)$ in the matching methodology? Do correlated X variables carry along their correlated Y and Z variables properly?

Distance Functions

How do the various dissimilarity measures affect the resultant merge file? What is a "correct" distance function? (See Kadane, 1975.) In practice, distance functions usually reflect the data aspects of greatest importance in the target microanalytic model or database.

For other research questions and issues surrounding merging activities, see the Federal Committee on Statistical Methodology (1980).

1.10 Experimentation Overview

To benchmark the various merging schemes and study the statistical aspects of the merge process, a series of experiments was performed with public-use national data sets. The 1975 SIE file was designated to be a test population from which a selected series of samples was drawn. Each record item was declared to be in set X, Y, or Z based on data type and correlations with other data items. The resultant sets of files were merged pairwise in various combinations using a variety of distance functions, merge schemes, and levels of data bias and error. By designating the full SIE file to be the population, the actual (X, Y, Z) is known, unlike the usual case in practice. This availability of the complete population provides an accurate standard for comparison with any merge file.

The experimental design was structured to study the effect, if any, of the above parameters on Y-Z relationships, standard statistical tests, and measures of "goodness" of the match. The study also investigated the sensitivity of the various merge algorithms to the distance function used and to the introduction of bias and error.

2. STATISTICAL FRAMEWORK AND EXPERIMENTAL DESIGN

2.1 Notation and Overview of the Study

Statistical matching methods have been developed to combine the information from two microdata files, each collected from a separate sample survey, into a single composite file. The objective of statistical matching is to create a single file that is equivalent to a valid sample taken from the population of interest.

The two input files, A and B, are of the form $(X1, Y)$ and $(X2, Z)$, respectively, where $(X1, Y)$ is a sample with multivariate observations (x_i, y_i) on each sampling unit, while $(X2, Z)$ is another independent sample with multivariate observations (x_j, z_j) on each sampling unit. Note that sets $X1$ and $X2$ are measured on the same data items (e.g., wages, interest income, family size), but the observation sets $X1$ and X2 are measured on different sampling units arising from the two different surveys. The data items $(X1, Y)$ and $(X2, Z)$ are obtained from either stratified or probability samples taken from the national population. The number of observations is typically large; e.g., in excess of 50,000 records.

From such files, a statistical match would create a single file of the form $(X12, Y, Z)$, where set $X12$ is a composite of $X1$ and $X2$. Presumably, such a file would be used in practice as if it were a valid random sample taken from the population of (X, Y, Z) measurements. Statistical inferences would be made with standard methods developed to account for sampling variability of such random samples.

A fundamental question to be addressed is: When do matched files really contain the same sampling variability as ordinary random samples? It is the goal of this project to empirically investigate the performance of some known matching methodologies from this point of view. The experimentation took the general form of (1) creating file A and file B from known populations, (2) statistically matching the two files, and (3) calculating a statistical summary of the matched files. By repeating these steps many times for each matching methodology, the empirical sampling results of the matched files may be compared with the known sampling properties of valid random samples.

2.2 Statistical Matching Issues to Be Addressed

Statistical Inference with Matched Files

One objective of this work is to identify conditions under which the matching methodologies will perform well. If a matching technique produces matched $(X12, Y, Z)$ files that behave like random samples (X, Y, Z), then the technique would be totally successful. This, however, may be too strict a requirement. A weaker condition for judging a matching technique to be acceptable would be to require that point estimates of the (X, Y, Z) population parameters be unbiased or consistent. Matched files that provided accurate estimates would be of great value, even if the precisions of such estimators were difficult to assess.

By conducting many replications of the matching process for known populations, it is possible to statistically study the properties of matched file estimators of key population parameters, such as $\mathrm{cov}(Y, Z)$. Since the success of a matching technique will quite possibly depend on the properties of the sampled population, the experiments were conducted for a variety of theoretical populations that are expected to affect the matched file estimators in different ways.

Constrained Versus Unconstrained Procedures

As described previously, matching procedures may be divided into two principal types, constrained and unconstrained. In each case, the $(X1, Y)$ file is linked, record by record, to the $(X2, Z)$ file to form the $(X12, Y, Z)$ composite file. In unconstrained matching, each record in the $(X1, Y)$ file is matched with the single closest record in the $(X2, Z)$ file. The composite $(x12, y, z)$ record weight is the weight of the $(x1, y)$ record. In constrained matching, each record in each file may be matched one or more times; however, for a given record, the sum of the linked record weights must equal the original weight.

The desirable statistical characteristic of unconstrained matching is that the degree of association between $X1$ and $X2$ is closer at the unit level than is the unit-level association of $X1$ and $X2$ in a constrained match. The potential disadvantage of unconstrained matches is that the statistical characteristics of $(X2, Z)$ can be altered in the matching process, assuming that the $(X2, Z)$ file is the one that is unconstrained. The advantage of constrained matching is that all statistical properties of $(X1, Y)$ and $(X2, Z)$ are preserved in the matching process. It must be noted that the statistical characteristics of $(X1, Z)$ might not be the same as those of $(X2, Z)$, even though the data item $X1$ can be accepted as statistically equivalent to $X2$. The disadvantage of constrained matching is that unit-level associations between $X1$ and $X2$ are not as close as can be obtained using unconstrained matching. However, for both constrained and unconstrained matching, the ultimate test is whether or not the matched file generated is statistically equivalent to a valid sample of (X, Y, Z) drawn from the population of such data items. The question becomes one of identifying the conditions under which either constrained or unconstrained matches produce valid results.

Cov(Y, Z|X)

If Y and Z are uncorrelated for given levels of X–i.e., $\text{cov}(Y, Z|X = x) = 0$ for all values x–then nonmatching methods could be used to estimate the joint distribution of (X, Y, Z) from information obtained from the unmatched files alone. As pointed out by Sims (1978), if, under conditional independence, the joint distributions of (X, Y, Z) admit probability density functions, then

$$f_{XYZ}(x, y, z) = f_{XY}(x, y) \ f_{XZ}(x, z)/f_X(x) \tag{6}$$

The probability density functions on the right-hand side of equation (6) could all be estimated from the separate $(X1, Y)$ and $(X2, Z)$ files, and from this the joint distribution of (X, Y, Z) could be estimated. Then, any population parameters of interest could be estimated using this estimated joint distribution. However, when the data files are large, the computational effort for this approach may be as great as or greater than that for matching techniques. Furthermore, statistical properties for this estimation approach might require unusual methods not available in standard statistical computing packages.

Current matching techniques usually accomplish the match by aligning $X1$ and $X2$ values that are close by some distance-function criterion. (See Section 2.3.) Since information about Y and Z is not used in the matching criteria, it would seem that the created matched files will likely have sample $\text{cov}(Y, Z|X = x)$ close to 0. This is because when several records have exactly the same x information, the matching is accomplished within these records by arbitrary or random selection. However, this might not be a problem if, in fact, there are only a few records with the same set of x values. Most matching projects have not included Y-Z relationships in the matching methodology, even though it is not assumed that $\text{cov}(Y, Z|X = x) = 0$. One of the major objectives of this project is to examine the results of matched files that do not use Y-Z relationships in the matching when, in fact, $\text{cov}(Y, Z|X = x) \neq 0$.

2.3 Distance Functions and Matching Methodologies

The matching methodologies considered here all proceed by defining a distance function that measures the dissimilarity between a pair of records. This function assigns a value d_{ij} to any pair of records (x_i, y_i) and (x_j, z_j)

from files A and B, respectively. For a given match, say, M, of the two files, an overall distance D_M is defined as a weighted sum of the distances of all matched record pairs as follows:

$$D_M = \sum_{(i,j)\in M} w_{ij} d_{ij} \tag{7}$$

In unconstrained and constrained matches, the final matched file is chosen as the match M^* that minimizes D_M over whatever class of possible matches is being considered.

This study focuses on four types of distance functions. There are two major groups, weighted absolute-difference methods and Mahalanobis distances. In addition, each of these types may be applied to the X items alone or expanded to include all X, Y, and Z items. Each of these four types of distance functions is used in conjunction with both an unconstrained and a constrained-optimal matching scheme, thus giving eight primary matching methodologies for study.

Weighted Absolute-Difference Measures

Distance functions in this category are of the type used by the U.S. Department of the Treasury's Office of Tax Analysis (Barr and Turner, 1980) and the Social Security Administration's Office of Research and Statistics (Radner, 1980). This function uses subjective weights, reflecting the relative importance of each data item, which are multiplied by the absolute differences of values of the corresponding items in the pair of records under consideration. Specifically, when only the X items are being considered, the distance between record i in file A and record j in file B is defined as:

$$d_{ij} = \sum_{k=1}^{r} s_k \cdot |x1_{ik} - x2_{jk}| \tag{8}$$

where $x1_{ik}$ and $x2_{jk}$ denote the kth data items in the respective files, r is the number of data items in X, and s_k is the subjective weight for data item k.

This procedure can be expanded to include additional ($X12$, Y, Z) relationships by adding other difference terms to the function definition. For example, to include some information about the relation of data item k of X and data item λ of Y, an additional component of the distance function could be $s_{\lambda k} \cdot |y1_{i\lambda} - x2_{jk}|$. The weight s_k would be determined subjectively, with the sign of the term corresponding to the sign of $\text{cov}(X_k, Y_k)$. Similarly, relationships between the various X items themselves, X and Z items, and even Y and Z items could be included in the matching criteria. Of course, the choice of the subjective weights is important, since they will have a strong impact on the matches obtained.

Mahalanobis Distance Metrics

The other category of distance functions studied was proposed by Kadane (1978). One procedure, which uses only the X items, defines

$$d_{ij} = (s1_i - x2_j)'(\Sigma_{xx})^{-1}(x1_i - x2_j) \tag{9}$$

where Σ_{xx} is the covariance matrix of the X variables. This is the Mahalanobis distance between two x values and, using only the X information, it arises as the maximum-likelihood solution for exact matching of normal random variables. It seems quite plausible that it will also perform well in statistical matching.

Kadane also suggests a procedure that employs full (X, Y, Z) information. In this instance, file A is expanded from (X, Y) to $V \equiv (X1, Y, \hat{Z})$, where $\hat{z}_i \equiv E(Z|X = x1_i, Y = y_i)$ is the regression prediction for the missing Z data of record i. Likewise, file B is expanded to $U \equiv (X2, \hat{Y}, Z)$, where $\hat{y}_j \equiv E(Y|X = x2_j, Z = z_j)$. S_1 and S_2, the covariance matrices of $(X1, Y, Z)$ and $(X2, Y, Z)$, respectively, may be formed from Σ, the covariance matrix of (X, Y, Z). The match is performed using

$$d_{ij} = (v_i - u_j)'(S_1 + S_2)^{-1}(v_i - u_j) \tag{10}$$

A difficulty here is in obtaining accurate Σ_{XX} entries used in calculating S_1 and S_2. These must come from a source outside the two files being matched or, if $\text{cov}(Y, Z|X = x) = 0$ is assumed, they can be calculated as

$$\Sigma_{YZ} = \Sigma_{YX}(\Sigma_{XX})^{-1}\Sigma_{XZ} \tag{11}$$

2.4 Experimental Design

To gain insight into the impacts of merging scheme, distance functions, and measurement error in the data, a set of experiments was performed using national data sets.

The 1975 Survey of Income and Education was treated as a population and, from this file, five randomly drawn samples of 1000 records each were drawn. The records of each file were divided into two new data sets by designating each record item to be in set X, Y, or Z and forming an (X, Y) file and an (X, Z) file. These files were merged using various methodologies and examined using the evaluation design described in the section that follows.

The data items designated for sets X, Y, and Z were selected to include each of the various types of data available on the file and different levels of correlation. Also, the files were merged using different numbers of X variables. The performance of the matching methodologies can be simultaneously evaluated, by comparison with known characteristics of the original records.

The existence of measurement error is simulated by adding bias, unbiased noise, and biased noise to subsets of the X variables prior to merging. The sensitivity of the distance-function definitions and merging schemes to such error is then evaluated both at the record level and in the aggregate using the statistical evaluation design.

2.5 Statistical Evaluation of the Merged Files

Bias for Matched File Estimators

Does a matching methodology induce a bias in the estimation of important population parameters? To answer this question, the key parameters of interest will be cov(Y, Z) and cov(X, Z) with μ_Z and var(Z) of particular interest for unconstrained matching. Let Θ be a parameter of interest. Note that Θ will be known exactly, since we know the population from which we have samples. Also, let T_i be the estimate of Θ derived from the ith matched file generated with some particular methodology. Now, let μ_T be the mean of all T that could possibly have been obtained as estimates of Θ from matched files. Then, the matching procedure produces unbiased estimates of Θ if $\mu_T = \Theta$.

The question of biased estimators of Θ can then be addressed by comparing the observed T_i with Θ. A simple test for $\mu_T = \Theta$ can be made by using

$$z = \frac{\sqrt{k}(\bar{T} - \Theta)}{S} \tag{12}$$

where

$$\bar{T} \equiv \frac{\Sigma_{i=1}^{k} T_i}{k} \tag{13}$$

and

$$S^2 \equiv \frac{\Sigma_{i=1}^{k} (T_i - \bar{T})^2}{k-1} \tag{14}$$

For large values of k (including this study's $k = 100$), z will have an approximate standard normal distribution if $\mu_T = \Theta$.

As noted earlier, a randomized block-design analysis would be available for comparing the $T_i - \Theta$ biases of several methodologies simultaneously.

Matched Samples Versus Valid Random Samples

Does the sampling variability of matched files resemble the variability in random samples? Here, the major concern is with the overall sampling distribution of the (X, Y, Z) merged files or simply with the bivariate sampling distribution of (Y, Z) obtained from merged files.

An excellent way to examine a multivariate sample is to divide each variable into classes and form a multiway contingency table of the sample. For example, for a (Y, Z) sample and selected values of a, b, c, d, e, and g, we can analyze the table below, where f_{ij} is the observed frequency of data in cell (i, j).

	$z \le d$	$d < z \le e$	$e < z \le g$	$g < z$
$y \le a$	f_{11}	f_{12}	f_{13}	f_{14}
$a < y \le b$	f_{21}	f_{22}	f_{23}	f_{24}
$b < y \le c$	f_{31}	f_{32}	f_{33}	f_{34}
$c < y$	f_{41}	f_{42}	f_{43}	f_{44}

Since the (Y, Z) population is known, the expected value for each f_{ij} may be calculated for random sampling. It is well known that the Pearson chi-square statistic

$$\chi^2 \equiv \sum \frac{[f_{ij} - E(f_{ij})]^2}{E(f_{ij})}$$

has an approximate chi-square distribution with 15 degrees of freedom when valid random sampling is done. This suggests that we calculate χ^2 for each matched file that we obtain using a given methodology; if the matched files are equivalent to random sampling, then these k χ^2 values should follow the appropriate chi-square distribution. The Kolmogorov test can be used to test this, or we can simply describe χ^2 values with a histogram. A similar procedure can be used to compare the sampling variability of whole matched files with that of valid random sampling.

If we wish to bypass the file-by-file comparison of matched file distribution and random sampling, all matched file data can be pooled for a given methodology into a single table and the pooled distribution compared with the expected values calculated from the known population. To test the hypothesis that the matched file samples were simple random samples, a single Pearson chi-square statistic can be used.

3. ANALYSIS BASED ON A NATIONAL MICRODATA SET

The objective of this section is to empirically examine the statistical characteristics of selected matched files, given that the population characteristics of the data are known. These comparisons will be used to evaluate the matching techniques.

The primary statistical procedure employed in these comparisons is a simple χ^2 goodness-of-fit test using percentage distributions and prespecified categories. In addition, matched file correlation matrices are compared directly with the population correlation matrix. No specific test for equality of correlation matrices was applied, since direct observation of the correlation matrices indicated that, for certain data items, there existed very large deviations.

An absolute-difference distance function was tested under conditions of noise, bias, and combined noise and bias. The same function was studied for sensitivity to the number of common variables used, and the Mahalanobis distance function was also tested. Also investigated were the consequences of matching a sample with itself under a wide variety of circumstances and the effect of using an unconstrained merging procedure instead of a constrained model.

This work focuses on the $Y-Z$ distributions in the matched files and their relationship to the population Y-Z distributions, since the implied motivation for most statistical matching is the construction of Y-Z distributions. Before this study, published merge analyses paid almost exclusive attention to the relationship between $X1$ and $X2$ in the matched file and to the vector of Z means. However, in these previous studies, data limitations have been such that ex-post-facto testing of *Y-Z* distributions has not been possible.

Fifty matched files were generated and solved with the transportation model for empirical analysis of the issues mentioned above. These constrained matches employed five samples of approximately 1000 records each, selected from the SIE family file. In addition to the constrained matches, a number of unconstrained matches were generated for comparison purposes.

3.1 Population, Samples, and Data Item Descriptions

In this phase of the study, an extract of the SIE family file, resident in the Office of Tax Analysis' data library, was designated to be a base population. This particular file has 54,034 records representing a full population of approximately 78 million families. The data items selected for inclusion in the extract file were family wage or salary income, interest

Table 3. Files Created from Perturbations of SIE5

Name	Modification(s) Made to Record Item(s)
SIE6A	Asset value was reduced by 20% for simulated bias.
SIE6B	Asset value was reduced by 10% for simulated bias.
SIE7A	25% of the records have asset value multiplied by a random number between 0.75 and 1.25 for simulated noise.
SIE7B	All asset values were multiplied by a random number between 0.75 and 1.25 for simulated noise.
SIE7C	25% of the records have asset value multiplied by a random number between 0.9 and 1.1 for simulated noise.
SIE7D	All asset values were multiplied by a random number between 0.9 and 1.1 for simulated noise.
SIE8A	25% of the records have asset value multiplied by a random number between 0.7 and 1.0 for simulated noise and downward bias.
SIE8B	All asset values were multiplied by a random number between 0.7 and 1.0 for simulated downward bias.

NOTE: Samples SIE6A through SIE8B are otherwise the same as SIE5, and different random numbers were used for each randomly perturbed record.

income, countable assets, age of the family head, highest grade of school completed by the family head, sex of the family head, total annual family income, social-security income, number of adults in the family, dividend income, family size, and race of the family head.

Five subsamples of approximately 1000 records each were randomly selected for matching purposes. In addition, one of the subsamples (SIE5) was used to form "other samples," with the data items perturbed to simulate noise and bias.

The five subsamples' identifiers and their respective sizes are: SIE1, 938 records; SIE2, 943 records; SIE3, 942 records; SIE4, 991 records; and SIE5, 951 records. The weights of each file summed to approximately 1.4 million. The subsamples created from SIE5 and their characteristics are described in Table 3.

Six items were designated as *X*, or common, variables: wages and salaries, interest income, assets, age of family head, highest grade of head, and sex of family head. Total income, social-security income, and number of adults were chosen to be the *Y* variables (i.e., the variables unique to the first matching file, A). The variables selected to be set *Z* (i.e., the variables unique to the second matching file, B) were dividend income, family size, and race.

The *Y-Z* correlation matrix for the full population and the differences between each subsample's *Y-Z* correlation matrix and the population correlations are given in Tables 4 through 9. In addition, the population percentage frequency counts for all *Y-Z* items pairs are given in Tables 10 through 18.

3.2 Matched Files Generated Using the Transportation Model

Fifty matched files were generated using the Office of Tax Analysis' optimal-constrained merge system (see Barr and Turner, 1981, for de-

Table 4. Full SIE Population *Y-Z* Correlation Matrix

Total Income	1.00					
Social Security	−.14	1.00				
Number of Adults	.44	−.04	1.00			
Dividends	.33	.06	.02	1.00		
Family Size	.34	−.17	.76	−.01	1.00	
Race	−.12	−.15	−.02	−.04	.05	1.00

Table 5. Difference Between SIE1 Correlation Matrix and Population Correlation Matrix

Total Income	0					
Social Security	.05	0				
Number of Adults	.06	.01	0			
Dividends	−.03	.11	.05	0		
Family Size	.03	−.17	.02	−.01	0	
Race	−.12	.08	.02	−.04	.05	

Table 6. Difference Between SIE2 Correlation Matrix and Population Correlation Matrix

Total Income	0					
Social Security	−.02	0				
Number of Adults	−.01	−.03	0			
Dividends	−.14	−.01	−.03	0		
Family Size	−.04	.03	−.04	−.01	0	
Race	−.04	−.01	−.02	−.04	.02	0

Table 7. Difference Between SIE3 Correlation Matrix and Population Correlation Matrix

Total Income	0					
Social Security	.01	0				
Number of Adults	.01	−.05	0			
Dividends	−.05	.14	.02	0		
Family Size	.05	−.04	.03	−.01	0	
Race	−.08	−.07	.09	−.01	.11	0

Table 8. Difference Between SIE4 Correlation Matrix and Population Correlation Matrix

Total Income	0					
Social Security	.03	0				
Number of Adults	−.03	−.05	0			
Dividends	.06	−.01	−.03	0		
Family Size	−.07	−.03	−.01	−.02	0	
Race	−.03	.06	−.06	.00	−.03	0

Table 9. Difference Between SIE5 Correlation Matrix and Population Correlation Matrix

Total Income	0					
Social Security	−.02	0				
Number of Adults	−.01	.02	0			
Dividends	.02	−.03	−.02	0		
Family Size	.03	−.01	.02	−.01	0	
Race	−.08	.01	.00	−.01	−.03	0

Table 10. Population Total Income and Dividend Joint Distribution (Percentage Counts)

	Dividends		
Total Income	$0	$1 under $1000	$1000 Plus
Under $5000	21.01	1.18	0.16
$5000 under $10,000	20.20	1.85	0.58
$10,000 under $15,000	16.75	2.24	0.55
$15,000 under $20,000	11.61	2.22	0.56
$20,000 under $25,000	6.74	2.00	0.54
$25,000 Plus	6.84	3.09	1.78

Table 11. Population Total Income and Family Size Joint Distribution (Percentage Counts)

	Family Size			
Total Income	1	2	3	4 Plus
Under $5000	13.48	4.85	1.93	2.09
$5000 under $10,000	7.69	7.71	2.97	4.27
$10,000 under $15,000	3.57	6.23	3.72	6.05
$15,000 under $20,000	1.20	4.06	2.91	6.22
$20,000 under $25,000	0.45	2.65	1.92	4.29
$25,000 Plus	0.46	2.89	2.35	6.06

Table 12. Population Total Income and Race Joint Distribution (Percentage Counts)

	Race	
Total Income	White	Nonwhite
Under $5000	17.82	4.53
$5000 under $10,000	19.52	3.11
$10,000 under $15,000	17.62	1.94
$15,000 under $20,000	13.25	1.13
$20,000 under $25,000	8.70	0.61
$25,000 Plus	11.08	0.67

Table 13. Population Social Security and Dividend Joint Distribution (Percentage Counts)

	Dividends		
Social Security	$0	$1 under $1000	$1000 Plus
$0	62.33	9.44	2.38
$1 under $3000	12.92	1.59	0.79
$3000 Plus	7.9	1.55	1.01

Table 14. Population Social Security and Family Size Joint Distribution (Percentage Counts)

	Family Size			
Social Security	1	2	3	4
$0	17.26	17.42	13.35	26.20
$1 under $3000	7.83	4.60	1.31	1.56
$3000 Plus	1.77	6.35	1.14	1.21

Table 15. Population Social Security and Race Joint Distribution (Percentage Counts)

	Race	
Social Security	White	Nonwhite
$0	64.94	9.28
$1 under $3000	13.36	1.94
$3000 Plus	9.69	0.77

Table 16. Population Number of Adults and Dividend Joint Distribution (Percentage Counts)

	Dividends		
Number of Adults	$0	$1 under $1000	$1000 Plus
1	27.35	2.64	1.02
2	37.38	6.28	2.06
3 Plus	18.18	3.66	1.09

Table 17. Population Number of Adults and Family Size Joint Distribution (Percentage Counts)

	Family Size			
Number of Adults	1	2	3	4 Plus
1	26.62	2.05	1.41	0.95
2	0	26.33	7.43	12.02
3 Plus	0	0	6.96	16.00

Table 18. Population Number of Adults and Race Joint Distribution (Percentage Counts)

	Race	
Number of Adults	White	Nonwhite
1	26.05	4.96
2	41.55	4.23
3 Plus	20.22	2.74

scription). The subsamples designated in the previous section were selectively matched pairwise using six different distance functions. These six distance models use the X vector of common data items.

Weighted Absolute Differences Model

Model 1 is an absolute-difference distance function where, for record i from the first file and record j from the second file

$$d_{ij} = C_1 + C_2 + C_3 + C_4 + C_5 + C_6$$

where the six components are calculated as follows and the first component in the distance function for any given record match is

$$C_1 = \min\left\{400,\ 100\,\frac{|(\text{File A wage} - \text{File B wage})|}{\text{File A wage}}\right\}$$

The index C_1 is the absolute value of difference in wages and salaries for any pair of A and B records divided by the File A wage, but constrained not to exceed 400. For example, if the File A wages are \$25,000 and the File B wages are \$25,596,

$$C_1 = \min\{400, 100|(25{,}000 - 25{,}596)|/25{,}000\} = 2.4$$

In this example, the index C_1 denotes the fact that the given B record has a wage that differs from that of the A record by 2.4 percent. The upper limit of $C_1 = 400$ is arbitrary, but is intended to not allow differences in wages alone to determine a match for situations with large total distance; i.e., in excess of 400.

The record distance-function component, C_2, is a penalty assessed for differences in countable assets and follows the same formula as C_1.

The index C_3 denotes an index for differences in interest income between a pair of A and B records. Interval categories are used for the calculation of C_3, as defined in Table 19. The index C_3 has an upper limit of 52, which means that the greatest difference in property incomes has a distance-function penalty equivalent to a 52 percent difference in wages. Hence, the matching algorithm will try to maintain compatibility between the broad categories of property income, but the penalty for noncompatibility is never very large. In the lower segment of the income distribution, the impact of C_3 is to match records with zero property income together; whereas in the upper range of the income distribution, the index C_3 will keep records with large amounts of property income together, all else being equal.

Demographic factors included in the distance function are age, sex, and highest grade attained by head of household. The age penalty is defined by the variable C_4, which is described in Table 20. The age penalty is based on the age of the first person in the tax record.

The penalty for age difference is never greater than an 80 percent difference in wages. The broad age categories are defined to represent school age and retirement age. For example, the age break point of 62 represents early retirement and 66 denotes regular retirement. The age "17 and less" represents children living at home, and the age interval "18–22" represents college or beginning employment age. For the objectives of this matched file, persons with ages between 23 and 61 are not considered to be different, if all other factors are the same. However, a large penalty is imposed if a person 61 or younger is matched with a person 62 or older, in order to differentiate persons eligible for Social Security income from those who are not.

Table 19. C_2 = Interest Income Difference Index

	File B Interest Income				
File A Interest	$0	$1–1000	$1001–10000	$10,001–100,000	100,001 or more
0	0	13	26	39	52
$1–$1000	13	0	13	26	39
$1001–$10,000	26	13	0	13	26
$10,001–$100,000	39	26	13	0	13
$100,001 and above	52	39	26	13	0

Table 20. C_4 = Penalty Index for Difference in Ages

File A Age	File B Age				
	$\leqslant 17$	$18 \leqslant 22$	$23 \leqslant 61$	$62 \leqslant 65$	65 and over
$\leqslant 17$	0	12	32	80	80
$18 \leqslant 22$	12	0	24	80	80
$23 \leqslant 61$	32	24	0	64	80
$62 \leqslant 65$	80	80	64	0	40
66 and over	80	80	80	40	0

The penalty index for differences in highest grade attained by head of household is calculated as follows:

$$C_5 = 16 \cdot \mathrm{INT}(|\text{Grade of A} - \text{Grade of B}|/3)$$

where INT(x) is a function whose value is the smallest integer less than or equal to x. This value never exceeds a 100 percent difference in wages and represents a graduated penalty for increased differences in highest grade attained. Note that there is no penalty for a difference of under 3 years, a penalty of 16 for a 3- to 5-year difference, and so on.

The last penalty included in the distance function is the index C_6 for differences in sex of head of household. If the A record and the B record have different sex codes, then the index C_6 is set equal to 100, which has the same impact as a 100 percent difference in wages:

$$C_6 = \begin{cases} 0 \text{ if A and B have the same sex code} \\ 100 \text{ if A and B have different sex codes} \end{cases}$$

The distance-function value for a given potential record match is the sum of variables C_k. More precisely, the notation for the variable C_k discussed above should be C_{ijk}, where i denotes the ith A record, j denotes the jth B record, and k denotes the index for income and demographic characteristics. Hence,

$$\begin{aligned} d_{ij} &= \text{distance-function value for the } i\text{th A record} \\ &\quad\ \text{and the } j\text{th B record} \\ &= \sum_{k=1}^{6} C_{ijk} \end{aligned}$$

The objective of the distance function is to try to force matches within the intervals defined for interest income, age, highest grade, and sex, and to try to obtain very close absolute agreement based on wages and assets.

Mahalanobis Distance Model

Model 2 is the Mahalanobis distance-function value for record i from the first file (A), and record j from the second file (B) is defined in (9) as:

$$d_{ij} = (x1_i - x2_j)'(\Sigma_{xx})^{-1}(x1_i - x2_j)$$

where $x1_i$ is the vector of common data items from record i of file A
$x2_j$ is the vector of common data items from record j of file B
Σ_{xx} is the covariance matrix of the X variable from the population file

Other Constrained Models

Model 3 is Model 1 without *assets* in the distance function, and *Model 4* is Model 1 without *assets, age*, and *sex* in the distance function. *Model 5* is an absolute-value percentage-difference distance function using only *wages and salaries*; i.e., using only C_1 from Model 1. *Model* 6 uses only the *age* and *highest grade attended*, i.e., C_4 and C_5, of Model 1.

Matched Files Created

The specifications of the 50 generated matched files using the transportation algorithm are given in Table 21. For matching purposes, the Y elements of the file A samples and the Z elements of the file B samples were ignored.

These 50 matched files are in the following test classifications.

Matched files 1–10: Pairwise matching of all samples using an absolute-difference distance function (Model 1)
Matched files 11–20: Pairwise matching of all samples using the Mahalanobis distance function (Model 2)
Matched files 23–30: Matching the sample SIE5 with itself under conditions of noise, bias, and combined noise and bias (Model 1)
Matched file 21: Matching SIE5 with itself using the absolute-difference distance function and six common variables (Model 1)

Matched file 22: Matching SIE5 with itself using the Mahalanobis distance function (Model 2)

Matched files 31 – 38: Matching samples SIE1 with SIE5 under conditions of noise, bias, and combined noise and bias (Model 1)

Matched files 40 – 43: Pairwise matching of sample SIE5 with samples SIE1, SIE2, SIE3, and SIE4 using the absolute-difference distance function with five common variables (Model 3)

Matched files 44–47: Pairwise matching of sample SIE5 with samples SIE1, SIE2, SIE3, and SIE4 using the absolute-difference distance function with only three common variables (Model 4)

Matched file 39: Sample SIE5 matched with itself using the absolute-difference distance function with five common variables (Model 3)

Matched file 48: Sample SIE5 matched with itself using the absolute-difference distance function with three common variables (Model 4)

Matched file 49: Sample SIE5 matched with itself using the absolute-difference distance function with only the common variable *wages and salaries* (Model 5)

Matched file 50: Sample SIE5 matched with itself using the absolute-difference distance function with only the two common variables *age* and *highest grade attained* (Model 6)

Tests Used to Compare Matched File Distributions with the Population Distributions

Two tests were selected for comparing the *Y-Z* distributions in a matched file to calculate a simple χ^2 statistic for each *Y-Z* pair, based on cross-tabulated percentages for the categories specified in Tables 10 through 18 and using the population percentage counts from these tables as the expected values. The χ^2 statistic is calculated in the following manner. For a given cell k in a *Y-Z* table,

$$\chi_k^2 = \frac{(f_m - f_p)^2}{f_p}$$

where f_m = weighted percentage of matched records in the kth cell

f_p = weighted percentage of population records in the kth cell taken from Tables 10 through 18

$\chi^2 = \Sigma_{k-1}^{N} \chi_k^2$

N = number of cells in the *Y-Z* table

degrees of freedom = (number of rows)(number of columns) − 1

Also, for simplicity, cell counts with less than 1 percent of the cumulative frequency were set equal to 1, and all frequency counts were rounded to the nearest whole percent. This statistic was selected to neutralize the effect of having weighted samples with enormous cell values, where the slightest percentage difference will generate very large χ^2 figures. For example, a weighted sample with a weighted cell count of 1 million deviating by 1 percent from the population cell count would result in a cell χ^2 of 100, which would not pass a goodness-of-fit test with degrees of freedom less than or equal to 20.

Implied in the selection of a χ^2 goodness-of-fit test based on percentage distributions is the assumption that percentage counts are sufficient to represent the data. That is, for most applications using microdata, a cell percentage count of 20.5 percent is just as useful as knowing that the actual weighted frequency count is, for example, 287,000. Another important assumption for the goodness-of-fit test is that the appropriate cell-defining categories have been selected. For instance, if the categories for dividend income specified in Tables 10, 13, and 16 are sufficient for any use of dividend income, then χ^2 figures based on these categories are meaningful. However, it must be stated that for the purposes of this study, the categories were selected to have relevance with the restriction that low-count cells were avoided by aggregation. Consequently, for a small-frequency data item (such as *social security* used in Tables 13 through 15), the categories ($0, $1 to $2999, and $3000+) were selected so that cross-tabulated counts using the other variables are reasonable.

The second test used in the study is the direct comparison of the *Y-Z* correlation matrix of a matched file with the population *Y-Z* correlation matrix. The comparison is displayed by subtracting the population correlation matrix from the matched-file correlation matrix. Ideally, the matrix obtained would be zero or have all elements very close to zero; hence, the matched-file-generated *Y-Z* distribution is statistically the same as the corresponding population *Y-Z* distribution. As will be discussed later in the Chapter, a direct test for equality of correlation matrices is not necessary because of large differences observed between the matched-file and population statistics for *dividends* and *total income*, and *family size* and *number of adults*.

It must be noted that the matched file will be different to a certain extent from the population file, since the samples that are used for matching are slightly different from the population samples.

3.3 Comparison of Absolute-Difference and Mahalanobis Distance Functions

Matched files 1 through 20 identified in Table 21 can be used to compare matched files generated by an absolute-difference distance function and a Mahalanobis distance function. Matched files 1 through 10 represent all pairwise matchings of the five subsamples SIE1 through SIE5 selected from the population file using an absolute-difference distance function. Matched files 11 through 20 represent all pairwise matching of the five subsamples SIE1 through SIE5 using a Mahalanobis distance function.

Only the *Y-Z* distributions will be examined, since the transportation model leaves all original distributions in their original form. For example, the covariance matrix for *X*1-*Y* in the matched file is the same as the corresponding matrix in file A, and the covariance matrix for *X*2-*Z* in the matched file is identical to the corresponding matrix in file B.

Table 22 summarizes this χ^2 statistic for each of the nine frequency-count tables representing all *Y-Z* distributions. The rows in Table 22 were arranged to allow a direct comparison of Models 1 and 2 for the same input data files. For example, matched files 1 and 11 given in the first two rows of the table are for input data files SIE1 and SIE2, where matched file 1 uses Model 1 and matched file 2 uses Model 2. The row averages are for the average χ^2 for a given matched file for the nine *Y-Z* frequency tables. In all cases, the average for the matched file using Model 1 is less than for the matched file using Model 2 generated from the same input data files.

Table 23 illustrates one of the χ^2 calculations in Table 22. Table 23 gives the percentage counts of records for the *Y-Z* distribution *total income* and *dividend income* in matched file 1. An interesting feature of constrained matching models is that the marginal distributions in Table 23 are identical to the marginal distributions of the original files. For example, the marginal distribution of the *Y* variable *total income* in this table is identical to the marginal *Y* distribution in sample SIE1, and the marginal distribution of the *Z* variable *dividend* in this table is identical to the marginal *Z* distribution in sample SIE2.

The population percentages for the corresponding *Y-Z* distribution for Table 23 are given in Table 10 and, using the χ^2 figure previously defined, the resulting χ^2 is 1.3. With $(6)(3) - 1 = 17$ degrees of freedom, a χ^2 of 1.3 indicates that the distribution in Table 23 is, for all practical purposes, the same as the distribution in Table 10 and, consequently, for this *Y-Z* distribution, the matched file is the same as the population file.

Table 21. Matched Files Created Using the Transportation Matching Algorithm*

File A	File B SIE1	SIE2	SIE3	SIE4	SIE5	SIE6A	SIE6B	SIE7A	SIE7B	SIE7C	SIE7D	SIE8A	SIE8B
SIE1		#1[M1] #11[M2]	#5[M1] #15[M2]	#8[M1] #18[M2]	#10[M1] #20[M2]	#31[M1]	#32[M1]	#33[M1]	#34[M1]	#35[M1]	#36[M1]	#37[M1]	#38[M1]
SIE2	*		*	*	*								
SIE3	*	#2[M1] #12[M2]		*	*								
SIE4	*	#3[M1] #13[M2]	#6[M1] #16[M2]		*								
SIE5	#43[M3] #44[M4]	#4[M1] #14[M2] #42[M3] #45[M4]	#7[M1] #17[M2] #41[M3] #46[M4]	#9[M1] #19[M2] #40[M3] #47[M4]	#21[M1] #22[M2] #39[M3] #48[M4] #49[M5] #50[M6]	#23[MI]	#24[MI]	#25[MI]	#26[MI]	#27[MI]	#28[MI]	#29[MI]	#30[MI]

Legend: $\#_i[M_j] \equiv$ Matched file number i using model j, with File A and File B defined from row and column, respectively.
*Note: With this procedure, the designation of A and B is arbitrary; see B vs. A for missing A vs. B combinations.

Table 22. Contingency Table χ^2 Values Based on Population Percentages as Expected Values

Matched File	Total Income & Div.	Total Income & Family Size	Total Income & Race	S.S. & Div.	S.S. & Family Size	S.S. & Race	No. of Adults & Div.	No. of Adults & Family Size*	No. of Adults & Race	Row Average
1	1.3	4.0	2.2	1.5	2.4	.23	.78	37	.58	1.6
11	8.8	38.6	4.6	.7	13.8	1.0	13.0	306	3.75	10.5
2	5.3	7.3	3.0	1.2	6.8	.72	0.81	165	.56	3.2
12	10.2	48.3	6.1	1.9	18.8	1.6	8.25	376	5.1	12.5
3	2.8	7.5	1.5	1.7	3.0	1.1	1.31	133	1.03	2.5
13	13.4	45.9	5.2	2.3	9.1	1.96	8.5	268	3.63	11.2
4	4.5	4.3	1.8	0.8	2.9	.56	1.14	121	.86	2.1
14	12.0	48.6	3.7	2.2	16.1	1.5	10.0	336	2.74	11.8
5	1.3	6.3	3.2	1.8	3.4	.80	2.04	127	2.67	2.7
15	8.7	70.1	5.2	3.26	8.2	.58	7.67	387	3.78	13.4
6	2.3	6.2	7.4	3.5	2.3	1.7	2.67	154	1.93	3.5
16	8.8	75.8	4.0	2.83	11.0	1.96	5.97	331	3.46	14.2
7	1.4	8.2	3.2	1.9	3.2	2.0	1.72	127	1.03	2.8
17	6.2	67.3	4.2	3.6	14.6	2.4	6.75	405	1.87	13.4
8	2.9	9.3	.5	1.2	4.2	.13	1.97	134	1.43	2.7
18	11.2	66.0	3.6	2.67	20.5	.23	11.4	339	1.38	14.6
9	1.4	10.0	1.1	1.3	3.9	1.90	1.7	122	.71	2.8
19	8.4	66.2	3.6	3.6	16.7	1.80	6.32	348	2.81	13.7
10	1.8	5.8	.9	.81	1.3	.21	1.06	131	.58	1.6
20	14.2	63.3	2.6	1.7	14.9	1.58	7.34	378	3.7	13.7
Avg. for	2.5	6.9	2.5	1.6	3.3	0.9	1.5	125	1.1	
1-10	$\sigma=$ 1.4	$\sigma=$ 2.0	$\sigma=2.0$	$\sigma=0.8$	$\sigma=$ 1.5	$\sigma=0.7$	$\sigma=0.6$	$\sigma=$ 34	$\sigma=0.7$	
Avg. for	10.2	59.0	4.3	2.5	14.4	1.5	8.5	347	3.2	
11-20	$\sigma=$ 2.5	$\sigma=12.5$	$\sigma=1.0$	$\sigma=0.9$	$\sigma=$ 4.0	$\sigma=0.7$	$\sigma=2.3$	$\sigma=$ 41	$\sigma=1.1$	
Degrees of Freedom	17	23	11	8	11	5	8	11	5	

*Number of Adults and Family Size omitted.

Table 23. Matched File 1 Total Income and Dividend Joint Distribution (Percentage Counts)

	Dividends		
Total Income	$0	$1-under $1000	$1000 Plus
Under $5000	20.5	1.42	0.39
$5000 under $10,000	17.27	2.31	0.73
$10,000 under $15,000	17.47	2.34	0.43
$15,000 under $20,000	11.91	2.40	0.72
$20,000 under $25,000	7.30	2.32	0.43
$25,000 Plus	6.79	4.07	0.61

However, this result is true only if the relevant categories are those in Tables 10 and 23 and percentage distributions are sufficient for the data being represented.

The degrees of freedom for the *Y-Z* distributions are given in the bottom row of Table 22. It is observed in the column for *total income* and *race* that the average χ^2 for matched files 1 through 10 is 2.5, with a standard deviation of 2.0. A rough interpretation of these figures is that the mean χ^2 plus two standard deviations $= 2.5 + 4 = 6.5$, which is an acceptable χ^2 figure given $DF = 11$. In fact, for matched files 1 through 10, the mean χ^2 plus two standard deviations yields an acceptable χ^2 for all *Y-Z* distributions, with the exception of *number of adults* and *family size.*

The average χ^2 for the Mahalanobis distance function for each *Y-Z* distribution is given in the row averages for matched files 11 through 20. If two standard deviations are added to the mean χ^2 figures, the resultant sum is an acceptable χ^2 in only five of the *Y-Z* tables.

In summary, it is observed from Table 22 using the χ^2 test that the absolute-difference distance function is much better than the Mahalanobis distance function. It is also observed that at the 5 percent level of acceptance, all but one of the absolute-value distance function *Y-Z* distributions are acceptable.

Another way to compare matched files with the population file and to compare one matching model with another is to observe the difference between the correlation matrix of a matched file and the correlation matrix of the population file. Table 24 gives the result of subtracting the

Table 24. Average Correlation Matrix for Matched Files 1–10 Minus the Population Correlation Matrix (Absolute-Difference Distance Function)

Total Income	0					
Social Security	.02	0				
Number of Adults	.01	−.01	0			
Dividends	−.27	0	−.04	0		
Family Size	−.08	−.04	−.47	−.01	0	
Race	.02	0	−.01	−.01	.03	0

population correlation matrix given in Table 4 from the average correlation matrix obtained from matched files 1 through 10.

The blocked-in portion of Table 24 represents the *Y-Z* distribution where, for the ideal match, all entries should be zero or close to zero. It is observed from Table 24 that these are significant differences from zero, where the big differences are for the correlations between *total income* and *dividends* and between *number of adults* and *family size*. The difference between the population and the matched-file distributions for *family size* and *number of adults* was also very evident from Table 22. However, the difference between the population and matched-file distributions for *total income* and *dividends* was probably masked by classifying all dividends over $1000 in the class *$1000 Plus*. Another feature of the blocked portion of Table 24 is that six of the differences are negative and only one is positive, reflecting the fact that the matched-file correlations are, on the average, smaller in absolute value than are the population correlations.

Table 25 gives the difference between the average correlation matrix for matched files 11 through 20 and the population correlation matrix given in Table 4.

From Table 25, it is observed that the Mahalanobis distance function produces larger deviations from the population than does the absolute-difference distance function represented in Table 24. The blocked-in portion of Table 25 represents the *Y-Z* distributions, and it is observed that correlations between *family size* and *number of adults*, between *dividends* and *total income*, and between *family size* and *total income* are very different from the population correlations. As observed in Table 24, there is a strong tendency for the matched-file *Y-Z* correlations to be smaller in absolute value than the population correlations. It is also

Table 25. Average Correlation Matrix for Matched Files 11–20 Minus the Population Correlation Matrix (Mahalanobis Distance Function)

Total Income	0					
Social Security	.02	0				
Number of Adults	.01	−.01	0			
Dividends	−.32	−.08	−.05	0		
Family Size	−.40	.12	−.77	.03	0	
Race	.12	.07	.08	−.01	.03	0

observed from Tables 24 and 25 that the absolute-value distance function is better than the Mahalanobis distance function.

3.4 Comparison of Matched Files Generated with an Absolute-Value Distance Function Using a Range of Common Variables

Earlier in this chapter, matching Models 1, 3, and 4 were specified. Essentially, Model 3 is the same as Model 1 with the data item *assets* left out. In the population file, *assets* is strongly correlated (.57) with the common variable *interest*, moderately correlated (.38) with the *Y* variable *total income*, and highly correlated (.70) with the *Z* variable *dividends*. Model 4 is the same as Model 3 with the common data items *age* and *sex* left out. In the population file, age is moderately correlated (−.22) with the common variable *wages and salaries*, moderately correlated (−.34) with the common variable *highest grade of family head*, strongly correlated (.58) with the *Y* variable *social security*, and moderately correlated (−.20) with the *Z* variable *family size*.

The objective of this section is to compare Models 1, 3, and 4 to investigate the effect of altering the number of variables in the distance funtion. To achieve this objective, matched files 9, 40, and 47 are grouped together representing samples SIE5 and SIE4 matched, respectively, with matching Models 1, 3, and 4. Matched files 4, 42, and 45 are grouped together representing samples SIE5 and SIE2 matched, respectively, with Models 1, 3 and 4. Matched files 10, 43, and 44 are grouped together representing samples SIE1 and SIE5 matched, respectively, using matching Models 1, 3, and 4. Also, matched files 7, 41, and 46 representing files SIE5 and SIE3 are matched, respectively, using matching Models 1, 3, and 4.

Table 26 displays χ^2 statistics as defined previously to compare Models 1, 3, and 4 using pairwise matchings of sample SIE5 with samples SIE 1, SIE2, SIE3, and SIE4. From the table, it is observed from the row-average column that Model 4 yields the largest average χ^2 statistics in three of the four groupings. It is also observed from Table 26 that Model 4 has the largest column average in six of the nine frequency tables. Models 1 and 3 appear to generate matched files with the same overall differences from the population file.

Once again, it is very obvious that the *Y-Z* distribution for *number of adults* and *family size* is very poor, but the other distributions are reasonable, given that the samples are different from the population.

Models 1, 3, and 4 can also be examined using the average correlation matrices for matched files using the different models. The correlation results for Model 1 were given in Table 24. The differences between the average correlation matrix using Model 3 and the population correlation matrix are given in Table 27. The difference between the average correlation matrix using Model 4 and the population correlation matrix is given in Table 28.

The blocked-in portions of Tables 27 and 28 reflect the differences between the *Y-Z* distributions in the matched files using Models 3 and 4 and the population *Y-Z* distributions. As mentioned in the previous section, the ideal match would have zero or near-zero entries. However, as in the case for Model 1 observed in Table 22, there are significant differences in the matched correlations for *dividends* and *total income* and for *family size* and *number of adults*. These large differences are a result of the tendency for the matched-file correlations to have smaller absolute values than the absolute values of the corresponding correlations in the full population. The population correlation for *dividends* and *family income* is .33, as opposed to the average Model 3 corresponding correlation of .06 and the corresponding correlation from the average results from Model 4 of .05. For *family size* and *number of adults*, the population correlation is .76, the average Model 3 correlation is .31, and the average Model 4 correlation is .18.

It should be noted that the entries in Tables 24, 27, and 28 outside the blocked-in section are for the correlations within the *Y*s (given above the blocked-in portion) and for the correlations within the *Z* s (given to the right of the blocked-in portion). Any nonzero entries outside the blocked-in portions are a consequence of differences between the samples and the population, since the transportation algorithm forces the within-*Y* and within-*Z* correlations to be the same as the sample values.

Table 26. Contingency Table χ^2 Values Based on Population Percentages as Base

Matched File	Total Income & Div.	Total Income & Family Size	Total Income & Race	S.S. & Div.	S.S. & Family Size	S.S. & Race	No. of Adults & Div.	No. of Adults & Family Size*	No. of Adults & Race	Row Average
9	1.4	10.0	1.1	1.3	3.9	1.9	1.7	122	.71	2.8
40	4.3	12.9	1.1	.66	5.1	1.8	1.6	133	.83	3.5
47	3.3	14.9	.91	.89	14.8	1.0	.62	232	1.0	4.7
4	4.5	4.3	1.8	.80	2.9	.56	1.1	121	.86	2.1
42	3.8	4.6	1.3	.67	1.1	.45	2.0	88	1.2	1.9
45	11.8	7.3	2.5	.90	6.1	.83	1.2	214	.7	3.9
10	1.8	5.8	.9	.81	1.3	.21	1.1	131	.58	4.2
43	7.5	6.2	1.6	1.40	1.7	.43	2.1	114	.45	2.7
44	4.0	11.9	.72	1.83	7.4	2.3	1.1	223	1.1	3.8
7	1.4	8.2	3.2	1.9	3.2	2.0	1.7	127	1.0	2.8
41	2.3	4.3	4.7	1.8	2.6	2.2	.56	106	1.0	2.4
46	4.1	11.4	4.4	2.3	6.5	.93	2.1	196	.4	4.0
Avg. for	2.3	7.1	1.8	1.2	2.8	1.2	1.4	125	.8	
9, 4, 10 & 7	$\sigma =$ 1.5	$\sigma =$ 2.5	$\sigma =$ 1.0	$\sigma =$ .5	$\sigma =$ 1.1	$\sigma =$ .92	$\sigma =$ .35	$\sigma =$ 46	$\sigma =$ .2	
Avg. for 40	4.5	4.0	2.2	= 1.1	2.6	1.2	1.6	110	.9	
42, 43 & 44	$\sigma =$ 2.0	$\sigma =$ 4.0	$\sigma =$ 1.7	$\sigma =$ .6	$\sigma =$ 1.8	$\sigma =$ .9	$\sigma =$ .7	$\sigma =$ 19	$\sigma =$ .3	
Avg. for 47	5.8	11.4	2.1	= 1.5	8.7	1.3	1.3	216	.8	
45, 44 & 46	$\sigma =$ 4.0	$\sigma =$ 3.1	$\sigma =$ 1.7	$\sigma =$ .7	$\sigma =$ 4.1	$\sigma =$ .7	$\sigma =$ .6	$\sigma =$ 15	$\sigma =$ .3	
Degrees of Freedom	17	23	11.	8	11	5	8	11	5	

*Number of Adults and Family Size omitted.

Table 27. Average Correlation Matrix for Matched Files 40-43 Minus the Population Correlation Matrix (Model 3 Distance Function)

Total Income	0					
Social Security	−.02	0				
Number of Adults	−.01	.02	0			
Dividends	−.27	.02	.02	0		
Family Size	−.07	.04	.45	0	0	
Race	.02	.02	.05	.01	.04	0

Table 28. Average Correlation Matrix for Matched Files 44-47 Minus the Population Correlation Matrix (Model 4 Distance Function)

Total Income	0					
Social Security	−.02	0				
Number of Adults	−.01	.02	0			
Dividends	−.25	.01	−.03	0		
Family Size	−.11	−.01	−.58	0	0	
Race	.04	.04	.01	−.01	.04	0

In summary, the results of this section indicate that Models 1 and 3 are better than Model 4. The implication for matching is that it is possible to have too few common variables in the distance function. However, because of the mixed results from Models 1 and 3, it cannot be stated that too many common variables can degrade the accuracy of a generated matched file.

3.5 Matching Under Conditions of Noise and Bias

Samples SIE1 and SIE5 were matched using Model 1, and the results of this match are designated as matched file 10 in Table 21. Earlier in this chapter, samples 6A through 8B were identified as versions of SIE5 with noise and bias injected into the X-variable *assets*. The purpose of this section is to compare matched file 10 with matched files 31 through 38, which are identified in Table 21. In all cases, matching Model 1 is used.

Table 29 displays the χ^2 statistic as defined in the previous two sections for comparing sample SIE1 matched with SIE5 when bias and

Table 29. χ^2 Statistacs for Bias and Noise Tests

Matched File	Total Income & Div.	Total Income & Family Size	Total Income & Race	S.S. & Div.	S.S. & Family Size	S.S. & Race	No. of Adults & Div.	No. of Adults & Family Size	No. of Adults & Race	Row Average*
10	1.8	5.8	.9	.81	1.3	.21	1.1	131	.58	1.6
31	2.3	6.4	1.1	.35	1.8	.21	.73	133	.51	1.7
32	1.5	8.4	1.6	.33	1.8	.21	1.1	133	.76	2.0
33	2.8	6.1	1.3	.75	1.3	.21	.73	137	.58	1.7
34	1.6	7.9	1.2	.25	1.5	.21	.81	112	.75	1.8
35	2.3	7.1	1.3	.75	1.3	.21	1.21	132	.58	1.8
36	2.8	6.9	1.2	.33	1.3	.21	1.6	138	.78	1.9
37	1.1	6.0	1.2	1.6	1.8	.21	.73	117	.58	1.7
38	2.1	7.9	.78	.25	1.8	.21	.73	134	.76	1.8
Avg. for 31-38	2.1 $c =$ .62	7.1 $\sigma =$ .9	1.2 $\sigma =$ .23	.58 $\sigma =$ .46	1.6 $\sigma =$ .25	.21 $\sigma = 0$	.95 $\sigma =$ .32	130 $\sigma = 9.6$	.66 $\sigma = .11$	
Degrees of Freedom	17	23	11	8	11	5	8	11	5	

*Number of Adults and Family Size omitted.

noise are injected into SIE5. Match file 10 is for the unaltered sample SIE5, matched file 31 has *assets* in SIE5 reduced by 20 percent, matched file 32 has *assets* in SIE5 reduced by 10 percent, matched file 33 has a 25 percent noise factor in *assets* in 25 percent of the records in SIE5, matched file 34 has a 25 percent noise factor in *assets* in all records in SIE5, matched file 35 has a 10 percent noise factor in *assets* in 25 percent of the records in SIE5, matched file 36 has a 10 percent noise factor in *assets* in all SIE5 records, matched file 37 has an average 15 percent downward bias and noise factor in *assets* in 25 percent of the records in SIE5, and matched file 38 has an average 15 percent downward bias and noise factor in *assets* for all records in SIE5.

In Table 29, it is observed that the row average for matched file 10 is slightly better than the row averages for matched files 31 through 38. It is also observed from the column averages of matched files 31 through 38, when compared with the row entries for matched file 10, that, on the average, χ^2 statistics for matched file 10 are better than those of the matched files 31 through 38.

Once again, as in the two previous sections, the *Y-Z* distribution for *number of adults* and *family size* are very poor, and most of the other distributions are reasonable. The empirical result taken from Table 29 is that moderate amounts of noise, bias, and combined noise and bias do not greatly affect the *Y-Z* distributions in matched files.

3.6 Results of Matching a Sample with Itself

The sample SIE5 was matched with itself under a variety of conditions; i.e., using all the matching models and using the conditions of bias, noise, and combined noise and bias. The files of particular interest are described in Table 21 as matched file 21 (Model 1), 22 (Model 2), 39 (Model 3), 48 (Model 4), 49 (Model 5), and 50 (Model 6) and matched files 23 through 30 (noise and bias tests).

Matched files 39, 48, 49, and 50 (generated under conditions of a reduced set of common variables) are unique in that they are identical with sample SIE5; that is, each sample record is matched with itself and with the matched weights equal to the record weights. There are 951 records in sample SIE5 and consequently, there are 951 records in matched files 39, 48, 49, and 50. The correlation matrix for each of these files is identical to the correlation matrix for SIE5.

Matched file 21 is slightly different from sample SIE5. However, this matched file has 974 records, which implies that all except 24 records have been matched with themselves. These 24 exceptions have been split and

Table 30. Correlation Difference Matrix for Matched File 21 Minus SIE5 (Model 1)

Total Income	0					
Social Security	0	0				
Number of Adults	0	0	0			
Dividends	0	0	0	0		
Family Size	0	−.001	−.004	0	0	
Race	−.001	.004	−.009	0	0	0

cross-matched with each other. The consequences of the mismatching of 24 records can be observed in Table 30, where it is seen that the correlations are almost identical with the exception of *race* and *family size*, which have an approximate difference of −.01. However, the percentage frequencies in the table for *race* and *family size* differ between SIE5 and the matched file by less than 0.1 percent.

Thus, the results of matching a file with itself are perfect using Models 3, 4, 5, and 6 and near perfect using Model 1. However, the results obtained using Model 2, the Mahalanobis distance function, are very poor by comparison. The difference between the correlation matrix from matched file 22 and that from SIE5 is given in Table 31.

The impact of bias and noise on sample data can be studied by comparing the characteristics of matched files 23 through 30. In all cases, the results are nearly identical to those of matched file 21, with correlation differences from matched file 21 less than .003. Each of these output files differ from each other only by less than 7 of the 951 records in SIE5, with

Table 31. Correlation Difference Matrix for Matched File 22 Minus SIE5

Total Income	0					
Social Security	0	0				
Number of Adults	0	0	0			
Dividends	.365	−.071	.039	0		
Family Size	−.49	.094	−.891	0	0	
Race	.173	.079	.071	0	0	0

the number of matched records ranging from 967 to 974. The conclusion is that limited amounts of bias, noise, and combined bias and noise do not affect a matched file.

3.7 Analysis of Unconstrained Procedures

To investigate the impact of unconstrained procedures on the resultant composite file, a single sample file, SIE2, was merged with each of the four remaining sample files using an unconstrained method with the absolute-difference distance function of Model 1. In each case, the weights for SIE2 are not constrained, and the other file is used as the base file. By observing the distributional statistics of a Z variable, the effects of dropping the weight constraints are demonstrated.

The means and standard deviations of the Z-variable *dividend income* are shown in Table 32 for all five sample files. Table 33 shows the same statistics for SIE2 when used as file B. Not only do the means vary, depending on the base file, but the standard deviations are also distorted, as much as ± 38 percent from the original values.

Noise and bias factors also influence the pattern of data in the $X2$ and Z variables, as illustrated in Table 34. Files SIE6A through SIE8B are identical to file SIE5, but with the $X2$ variables perturbed with noise or bias as described in Table 3. Table 34 demonstrates the impact of the X-data perturbations on the same fundamental statistics. The presence of bias or noise tends to decrease the Z variance and to distort even the means either upward or downward.

Table 32. Dividend Income Statistics for Sample Files

	Dividend Income	
Sample File	Mean	Standard Deviation
SIE1	$227	$1471
SIE2	194	1717
SIE3	350	1942
SIE4	353	2929
SIE5	292	1990

Table 33. SIE2 Dividend Income Statistics After Unconstrained Merging

		Dividend Income from SIE2			
File A	File B	Mean	Percent Deviation*	Standard Deviation	Percent Deviation*
SIE1	SIE2	$176	(−9.2)	$1057	(−38.4)
SIE3	SIE2	268	(+38.1)	1922	(+11.9)
SIE4	SIE2	146	(−24.7)	1454	(−15.3)
SIE5	SIE2	186	(−4.1)	1210	(−29.5)

*Percent deviations from original SIE2 values per Table 32.

Of course, all these statistical changes are a result of the implicit modification of the weights on file B by the unconstrained merge process. Such variations are in contrast with constrained procedures, which, as an integral part of the merge process, maintain the original (or equivalent) record weights and, hence, preserve all the $X2$ and Z data items and their interrelationships.

Table 34. Unconstrained Merges: Noise and Bias Tests

		Dividends of File B	
File A	File B	Mean	Standard Deviation
SIE1	SIE5	$220	$1713
SIE1	SIE6A	317	1557
SIE1	SIE6B	294	1433
SIE1	SIE7A	283	1418
SIE1	SIE7B	216	1028
SIE1	SIE7C	258	1371
SIE1	SIE7D	286	1475
SIE1	SIE8A	246	1149
SIE1	SIE8B	295	1992
SIE5 Original File		292	1990

In summary, this information seems to suggest that unconstrained approaches not only have difficulty maintaining the basic descriptive statistics of Z variables, but also are influenced by bias and noise in the X variables, two problems not encountered in constrained procedures.

3.8 Summary and Results of the Real Data Empirical Investigation

All the issues outlined in the introduction of this chapter were addressed using the 50 constrained matched files defined in Section 3.2 and the unconstrained matched files discussed in Section 3.7. The patterns of the results obtained indicate that sufficient observations have been generated for some general conclusions. These results are:

1. The absolute-difference distance function yields significantly better results than does the Mahalanobis distance function.
2. Noise and bias have a nominal effect on constrained matched files.
3. When a file is matched with itself using the transportation model with an absolute-difference distance function, the desired matching of records is produced even under conditions of bias, noise, and combined noise and bias.
4. All original statistical content in the input files is preserved with constrained matching. However, there is a tendency for the absolute value of correlations between the Y-Z items to be reduced from the population values.
5. The quality of a match is reduced if too few common variables are used in the distance function.
6. The absolute-difference distance function generated acceptable Y-Z distributions in seven of the nine Y-Z distributions specified, using the correlations in Table 4 and the percentage distribution functions and categories specified in Tables 10 through 18.

It is extremely interesting to note that the absolute-difference distance function used with the transportation model generated the unacceptable Y-Z distributions when $\text{cov}(Y, Z|X)$ was clearly nonzero.

Perhaps the most important conclusion is that the next applied research topic in this area should be the identification and development of a matching criterion that has the "known" Y-Z relationship included. That is, the matching function should exclude nonvalid Y-Z patterns while encouraging the valid ones. This conclusion is based on the empirical evidence that the transportation model using an absolute-difference distance function produces acceptable Y-Z distributions in most, but not all, situations. Also, it is reasonable to predict that in a proper environment with the necessary matching software and data, a matching function could be developed that would generate acceptable distributions for all Y-Z pairs.

The long-run implication of this conclusion is that statistical matching would be a very useful tool for data preparation in situations where, for example, every 5 years, population Y-Z characteristics are observed from a sample and, during the intervening years, file matching is done when the data are available only in $X1$-Y and $X2$-Z distributions. In this situation, the population Y-Z characteristics are used to match $X1$-Y and $X2$-Z files such that the matched file Y-Z distributions conform to expected patterns.

4. SUMMARY AND CONCLUSIONS

The objective of this study was to empirically measure the quality of statistical matching. To achieve this result, many statistical matches were generated, and the properties of these matches were observed and analyzed.

Statistical matching creates a composite microdata file from two original microdata files A and B. The records in the composite file are formed by appending the records from file B onto the records in file A. Statistical-matching algorithms identify the record weight to be associated with the record formed by appending the jth record from file B onto the ith record in file A. The statistical-matching procedures currently used in large-scale matching projects are described in Section 1. The statistical framework and the experimental design are given in Section 2, and the statistical results are presented in Section 3.

This study is one of the first research studies of its kind in this area: an in-depth computational work designed to achieve a foothold of understanding into the statistical nature of files formed by state-of-the-art merging techniques. The study views from dual aspects—theoretical and direct empirical—an important set of questions not answered by the literature. An enormous body of data has been created and analyzed; in the process, over 50 linear programming problems were solved with dimensions of up to 2000 constraints and 1 million variables.

Details of this research effort have been documented and, in this concluding section, several general conclusions suggested by these results are presented. These interrelated summary and conclusions are organized under the following headings: (1) the viability of merging, (2) choosing a merge technique and distance function, (3) the effects of data perturbations, (4) improvements needed in existing methods, and (5) future research directions.

4.1 The Viability of Merging

There are several instances suggested by this study in which specific statistical merging techniques perform well, and others where merging to accomplish certain goals is perhaps not advisable. The study focused on the various merge techniques' abilities (or lack thereof) to preserve known relationships between data items that came from and were unique to separate files. These relationships were expressed in the form of correlation and covariance statistics and cross-tabulations of pairs of such items. These are the so-called *Y-Z* relationships.

There is evidence to suggest that applications requiring that *Y-Z* relationships be preserved in modestly broad categories can be obtained with generally good results from a merge file created by the transportation model and the weighted absolute-differences distance function described in Section 3.2. While data categories such as *wages between $5000 and $10,000* and *age between* 20 and 30 *years* would be considered modestly broad, the categories *wages between $5000 and $5100* and *age of 21 years* would not. Therefore, many existing microsimulation models, such as the U.S. Treasury's Individual Tax Model, which do not have extremely strict requirements in this area, are well suited to the use of merged files.

This is not to suggest that there cannot be any problems with using such files or that improvements cannot be made in their construction. The empirical study using SIE data demonstrated the ability of merged files to create acceptable *Y-Z* data relationships in most cases, but also provided an excellent example of the creation of illogical relationships. Specifically, several records were created to form single-person families containing two adults. While such spurious results lead to reasonable concerns about merging, it is clear that these erroneous record matches could have been avoided easily by adding a penalty to the process distance function for each illogical match pair. The extension of this notion to less clear-cut cases is discussed below.

Another application of merged files is for correcting or expanding an existing file's item to account for underreporting. When, for example, items in a given file are not deemed sufficiently trustworthy, that file might be merged with another primarily to upgrade those particular items. If the files have many items in common, this use might be focused on either $X1$-$X2$ relationships or $X1$-Z relationships. Such relationships seem to be retained by merging.

In the instance where there is no relationship between the noncommon items, for a given set of values for the common items—i.e., $\text{cov}(Y, Z|X) = 0$—the Y-Z distribution is replicated well by merge techniques. Of course, this situation is not as useful as where there is such a relationship—$\text{cov}(Y, Z|X) \neq 0$—a case that current merging techniques have difficulty replicating. While it might be thought that all nonzero relationships are forced to zero by the merge techniques studied, the testing based on SIE data indicated that such relationships were only softened, not eliminated. This means that relationships that are not an explicit part of the merge procedure (e.g., in the distance function) tend to be attenuated through random pairings, but, on average, hold to a certain, although lesser, degree. Hence, the user of merge files should be cautioned about heavy reliance on them if a high degree of accuracy is required.

4.2 Choice of Merging Model

In summary, the study indicated that the best results can be obtained by applying an optimal constrained-merge model with an absolute-difference distance function. Testing with real data files verified the superiority of the constrained approach, but found the Mahalanobis distance function to yield extremely poor results, most likely a consequence of the presence of nonnormally distributed data.

The number of common variables used in the distance function was also shown to have a strong effect on the representativeness of matched files. As expected, more variables seemed better than fewer, perhaps due to the procedures' inability to distinguish between records when only a few variables are used.

4.3 Effects of Data Perturbations

A notable finding from the SIE data study was the robustness of the constrained-merge techniques when the variables used in a distance function are subjected to noise, bias, and both noise and bias. When the transportation model with an absolute-difference distance function was used to match a sample file with itself, 99 percent of all records were matched correctly, even under varying levels and types of noise and bias.

This lack of sensitivity to such prevalent conditions of sample survey data is a very positive result that enhances the attractiveness of merging schemes in general.

4.4 Improvements Needed in Existing Methods

It appears likely that current techniques, including the transportation model, could be improved with a modicum of additional research. For example, distance functions should be designed to account for known relationships among noncommon variables and, at the very least, to rule out illogical matches.

It is very clear from many aspects of the research that for merging methods to perform well in maintaining Y-Z relationships, the procedures must inject some measure of control over those relationships. In the simplest case, distance functions should associate heavy penalties with record matches that are illogical not only in terms of $X1$-$X2$ data configurations, but for Y-Z combinations as well.

In the more general case, the higher order statistical relationships between X, Y, and Z items should be incorporated into the merging procedures. Research to identify more sophisticated distance functions or matching schemes that directly address data characteristics such as nonnormality and $\text{cov}(Y, Z|X) \neq 0$ should be undertaken to counteract shortcomings inherent in procedures that are quite robust along other dimensions.

From a procedural point of view, it would be extremely useful for all merge-file users if aggregate statistics for noncommon variable pairs could be collected at regular (5- or 10-year) intervals in order to calibrate ongoing merge models. Such information could be collected piecemeal and at various times for subsequent construction and maintenance of these statistical mosaics. For example, the correlations between some item pairs probably would not change dramatically from year to year and would need to be updated or verified only over large time intervals. However, if such statistics were available, new merging schemes likely could be designed to incorporate them and perhaps eliminate the problems associated with $\text{cov}(Y, Z|X)$ significantly different from zero.

4.5 Further Research Directions

In addition to the research topics described above, it is felt that this work is only a starting point for research into the theory and practice of microdata file merging. In general, this line of research should be continued to (1) devise merge methodologies that are better able to capture the *Y-Z* relationships, (2) identify criteria to determine when a pair of files can be said to be mergeable, and (3) study the impact of merge technique at the model-output level, as opposed to the data-input level, to gauge models' sensitivities to data perturbations from this source.

A very different line of research would be to explore the applicability of new learning techniques to merging and imputation problems. For example, neural network models can be "trained," through exposure to a large set of examples, to identify highly nonlinear relationships between variables. A typical neural network model takes a set of input values and performs weighted computations to create a set of output values. The network is trained by repeatedly comparing desired outputs with computed ones and, if there is a difference, adjusting the model's internal weights. This is repeated until the model can adequately reproduce the desired outputs from a given set of input values. The ability of such a model to handle noisy data has made it useful in image-processing and pattern-recognition applications.

Such a model might be used to learn the interrelationships among variables from a sample, for use in massive imputation. The model might be trained with $(X1, Y)$ data, using the $X1$s as input variables and the Ys as desired output variables, so as to automatically uncover the X-Y interrelationships. The model could then process a second file of $(X2, Z)$ values, using the $X2$s as inputs, and use the computed Y outputs for $(X2, Z, Y)$ records. The research would involve not only model construction and training, but also a statistical analysis of the resulting imputed file and a comparison with merged model results.

Acknowledgments

We wish to acknowledge the strong support and cooperation of the following organizations: the Office of Evalution and Technical Analysis, Department of Health and Human Services; the Office of Research and Statistics, Social Security Administration; the Internal Revenue Service and Office of Tax Analysis, Department of the Treasury; and our home universities. This research was also supported by the Department of Defence under contract number MDA 903-86-C0182, the Air Force Office of Scientific Research under contract numbers AFOSR 83-0278 and AFOSR 87-0199, the Office of Naval Research under contract number N00014-87-K-0223, and ROME Air Development Center under contract number SCEEE PDP/86-75.

We especially thank Nelson McClung, Fritz Scheuren, Al Schwartz, Joan Turek-Brezina, and David Zalkind for lending their resources, expertise, and encouragement to this project. Their many contributions and suggestions greatly enhanced the research and this report.

APPENDIX: DETAILED DESCRIPTION OF A SMALL MATCHING PROBLEM USING CONSTRAINED AND UNCONSTRAINED METHODOLOGIES

It can be shown algebraically and understood intuitively that the statistical merging technique chosen directly affects the statistical structure of the resultant composite matched file. The objective of this appendix is to illustrate these effects using two small hypothetical data files.

An Example Matching Problem

The data files, called A and B, that we will use in our examples have three records and four records, respectively. These records are completely described in Figure A.1. Note that files A and B have some items that are in common and some that are not. The objective of merging would be to form a file of composite records, each containing items from both files, as depicted in Figure 1. As in all merging and matching techniques, the common items are used for identifying records with like attributes for matching purposes.

The tabulations given in Table A.1 show that the weight totals for each file are equal, indicating identically sized sample populations. If we let a_i be the weight of the ith record in file A and b_j be the weight of the jth record in file B, this property can be expressed mathematically as:

$$a_1 + a_2 + a_3 = b_1 + b_2 + b_3 + b_4 \tag{A.1}$$

Table A.1 also indicates that the weighted-item sums are slightly different, indicating reporting or sample variations.

Overview of the Matching Problem

We shall let w_{ij} represent the weight assigned to the composite record formed by merging record i of file A with record j of file B, with a zero value indicating that the records are not matched. Microdata file merging may be viewed as a problem of finding a set of nonnegative values for all w_{ij}s.

FILE A RECORDS:

Record Number	Record Weight	Schedule Code	Reported Adjusted Gross Income	Reported Deductions
1	1000	1	16,000	3,200
2	2000	1	12,000	2,300
3	500	2	20,000	4,000
			Common Items	*Non-common Items*

FILE B RECORDS:

Record Number	Record Weight	Schedule Code	Reported Adjusted Gross Income	Family Size	Transfer Income
1	1400	1	14,000	2	500
2	400	2	19,500	4	0
3	1500	1	11,000	3	3,000
4	200	2	17,000	2	0

Figure A.1. Example files A and B.

Table A.1. Item Tabulations for Example Files (Weighted)

Description	File A	File B
Population size	3500	3500
Schedule code = 1	3000	2900
Schedule code = 2	500	600
AGI, total ($Thousands)	50,000	47,300
Reported deductions, total	9800	n.a.
Family size, average	n.a.	2.65
Transfer income ($Thousands)	n.a.	5200

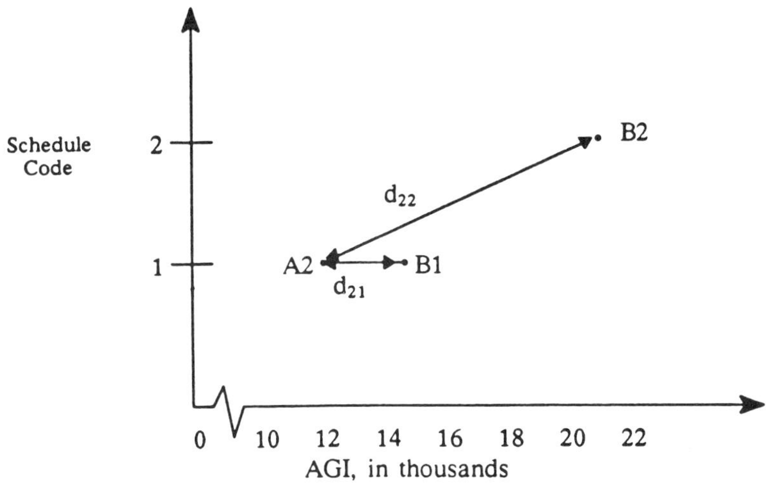

Figure A.2. Scatter diagram of selected records.

To guide the merge process to matching similar records a *distance function*, *d*, is used to measure the extent to which the attributes in any one record differ from the same attributes in another record. Intuitively, the parameter d_{ij} can be viewed as the "distance" between record *i* of file A and record *j* of file B, as illustrated in Figure A.2. In this example, file A record 2 (shown as point A2) is considered to be closer to file B record 1 (B1) than to file B record 2 (B2); that is, $d_{21} < d_{22}$, since the schedule codes and AGI values are in closer agreement.

A simplistic distance function will be used for illustrative purposes. (The effects of different dissimilarity metrics could also be studied using this example.) In this model, the interrecord distance will be defined as

$$d_{ij} = \frac{|(\text{File A AGI}_i) - (\text{File B AGI}_j)|}{100} + \begin{cases} 0, & \text{if schedule codes agree} \\ 25, & \text{if schedule codes differ} \end{cases}$$

Tableau A.1 can be used to summarize the matching problem. With a row for each record *i* in file A and a column *j* for each record in file B, a tableau cell (i, j) corresponds to a match possibility and a w_{ij} value. We will indicate a record match by including the composite record weight in a cell ($w_{ij} > 0$) and use a blank cell to mean that the records are not matched

FILE A RECORD	FILE B RECORD 1	2	3	4	a_i
1	20 w_{11}	50 w_{12}	50 w_{13}	35 w_{14}	1000
2	20 w_{21}	100 w_{22}	10 w_{23}	75 w_{24}	2000
3	85 w_{31}	5 w_{32}	115 w_{33}	30 w_{34}	500
	1400	400	1500	200	

Tableau A.1. Sample Tableau for Example Merge Problem

($w_{ij} = 0$). The box inset in each cell contains d_{ij}, the distance-function value. Row and column totals reflect the weights associated with the record.

Problem Constraints

Of course, any values could be assigned to the w_{ij} variables. However, since the record weights are an integral part of any computations made with the data items, these composite record weights directly affect the merge file's numerical structure. For this reason, we may wish the sum of the w_{ij} values for any record in file A to equal the original record weight, thereby not overmatching or undermatching that record and preserving that record's intrinsic data structure. In our example, this translates to the following set of constraints that we may wish to be in force in our solution to the merge problem:

$$
\begin{aligned}
w_{11} + w_{12} + w_{13} + w_{14} &= a_1 \\
w_{21} + w_{22} + w_{23} + w_{24} &= a_2 \\
w_{31} + w_{32} + w_{33} + w_{34} &= a_3
\end{aligned}
\tag{A.2}
$$

If, in addition, we wish to place the same conditions on the file B weights, we could also require:

$$\sum_{i=1}^{3} w_{ij} = b_j \qquad \text{for } j = 1, 2, 3, \text{ and } 4 \tag{A.3}$$

Since we assume that negative weights are not permitted, we always require the constraints:

$$w_{ij} \geqslant 0 \qquad i = 1, 2, 3 \text{ and } j = 1, 2, 3, 4 \tag{A.4}$$

Also, we may wish to use the distance-function values to achieve a best overall solution, so that our objective would be to require that the merge process:

$$\text{minimize } d_{11}w_{11} + d_{12}w_{12} + \cdots + d_{34}w_{34} \tag{A.5}$$

In so doing, we minimize the aggregate interrecord distance for the entire file.

Merging Techniques

Three statistical merging approaches will be considered in this study: unconstrained, constrained, and constrained-optimal. Each of these can be described in terms of some or all of the expressions (A.1) through (A.5).

The first, *unconstrained*, matching uses one file as a base and matches each record with the minimum-distance record in the other file. The merged records use the weights from the base-file records. This problem can be described mathematically by the set of expressions (A.1), (A.3), (A.4), (A.5) or (A.1), (A.2), (A.4), (A.5). In either case, one set of weight constraints is dropped from the problem.

Using our example files, one unconstrained match would be to drop the file-B constraints (A.3) and use file A as the base file. The solution that minimizes the total distance (A.5) is found by matching each file-A record with the minimum-distance file-B record. This solution is shown in Tableau A.2, where $w_{11} = 1000$, $w_{23} = 2000$, $w_{32} = 500$, and the remaining variables are zero. The result is a match with a low aggregate distance (42,500) and strong match statistics, but distortions in file-B data. By applying these record weights to the file-B data, as shown in the tabulations, note that not only are schedule code and AGI tabulations different, but the aggregate transfer income has increased by $1,300,000 and the average family size has grown from 2.65 to 2.94. Because the weights on file-A records are maintained by constraints (A.2), the tabulated values of these record items do not change.

FILE A RECORD	FILE B RECORD 1	2	3	4	
1	20 / 1000	50	50	35	1000
2	20	100	10 / 2000	75	2000
3	85	5 / 500	115	30	500
	1400	400	1500	200	

Total solution distance = 42,500

WEIGHTED TABULATIONS

Description	This Merged File	Original Value
File A Record Data:		
Schedule code =1	3,000	3,000
Schedule code =2	500	500
Total AGI (000s)	50,000	50,000
Total Deductions (000s)	9,800	9,800
File B Record Data:		
Schedule code =1	3,000	2,900
Schedule code =2	400	600
Total AGI (000s)	45,750	47,300
Transfer income (000s)	6,500	5,200
Avg. Family Size	2.94	2.65
Match Statistics (Wtd.)		
% Agreement on Schedule Code	100%	n.a.
Average Absolute AGI Difference	1,214	n.a.

Tableau A.2. Unconstrained File B (Ignore B Weights)

Tableau A.3 illustrates the case in which file-A weight constraints (A.2) are ignored, but file-B weight constraints (A.3) are enforced. The match solution for this situation is found by matching each file-B record with the closest record in file A. By using the file-B weights for merged records, the column totals are maintained, but the row weight totals are altered.

The result is, again, good match statistics and aggregate distance, but distorted data values, this time in the file-A items. Specifically, for file A, the schedule code tabulations are changed, total AGI has increased $2,400,000, and total deductions have increased $530,000.

Therefore, in either case, unconstrained matching can drastically distort the values associated with one file or the other. This is of particular concern in the case of the noncommon data. The purpose of matching is to be able to draw inferences regarding relationships between one data file and the noncommon items in another file. When the values from one file are distorted, the reliability of such inferences is lessened and, thus, the objective of matching is defeated.

An attempt to remedy this distortion problem is to include both sets of weight constraints, (A.2) and (A.3). This is called *constrained* matching and is described mathematically by expressions (A.1) through (A.4), with the imposition of (A.5) being the special case of *constrained-optimal* matching.

Tableau A.4 depicts a constrained nonoptimal match. Note that all match weights in a row sum to the row total (original file-A record weight), and column sums are kept similarly. By so doing, the original data structures are maintained and all tabulations are the same as those for the original files.

This improvement is not without its costs. The trade-off is in terms of total solution distance and poorer match statistics. The aggregate distance and average AGI discrepancy have more than doubled, plus schedule code agreement has dropped 20 percent, relative to the unconstrained matches.

To improve this solution to the greatest extent possible, expression (A.5) can be included and a constrained-optimal match sought. This optimization problem can be solved iteratively by devising a series of improved matches, each of which merges a new pair of records and drops an existing record match, while simultaneously maintaining the weight totals. Tableaus A.4 through A.6 illustrate this process.

Calculations from Tableau A.4 indicate that if the third record of A were matched with the second record of B ($w_{32} > 0$), w_{33} were set to zero, and adjustments were made to the other positive weights to maintain the

FILE A RECORD	FILE B RECORD 1	2	3	4	
1	[20] 1400	[50]	[50]	[35]	1000
2	[20]	[100]	[10] 1500	[75]	2000
3	[85]	[5] 400	[115]	[30] 200	500
	1400	400	1500	200	

Total solution distance = 51,000

WEIGHTED TABULATIONS

Description	This Merged File	Original Value
File A Record Data:		
Schedule code =1	2,900	3,000
Schedule code =2	600	500
Total AGI (000s)	52,400	50,000
Total Deductions (000s)	10,330	9,800
File B Record Data:		
Schedule code =1	2,900	2,900
Schedule code =2	600	600
Total AGI (000s)	47,300	47,300
Transfer income (000s)	5,200	5,200
Avg. Family Size	2.65	2.65
Match Statistics (Wtd.)		
% Agreement on Schedule Code	100%	n.a.
Average Absolute AGI Difference	1,457	n.a.

Tableau A.3. Unconstrained File A (Ignore A Weights)

FILE A RECORD	FILE B RECORD 1	2	3	4	
1	[20] 1000	[50]	[50]	[35]	1000
2	[20] 400	[100] 400	[10] 1200	[75]	2000
3	[85]	[5]	[115] 300	[30] 200	500
	1400	400	1500	200	

Total solution distance = 120,500

WEIGHTED TABULATIONS

Description	This Merged File	Original Value
File A Record Data:		
Schedule code =1		3,000
Schedule code =2	Same	500
Total AGI (000s)		50,000
Total Deductions (000s)	Values	9,800
File B Record Data:		
Schedule code =1	As	2,900
Schedule code =2		600
Total AGI (000s)	Original	47,300
Transfer income (000s)		5,200
Avg. Family Size		2.65
Match Statistics (Wtd.)		
% Agreement on Schedule Code	80%	n.a.
Average Absolute AGI Difference	2,942	n.a.

Tableau A.4. Constrained Match

FILE A RECORD	FILE B RECORD 1	2	3	4	
1	[20] 1000	[50]	[50]	[35]	1000
2	[20] 400	[100] 100	[10] 1500	[75]	2000
3	[85]	[5] 300	[115]	[30] 200	500
	1400	400	1500	200	

Total solution distance = 60,500

WEIGHTED TABULATIONS

Description	This Merged File	Original Value
File A Record Data:		
Schedule code =1		3,000
Schedule code =2	Same	500
Total AGI (000s)		50,000
Total Deductions (000s)	Values	9,800
File B Record Data:		
Schedule code =1	As	2,900
Schedule code =2		600
Total AGI (000s)	Original	47,300
Transfer income (000s)		5,200
Avg. Family Size		2.65
Match Statistics (Wtd.)		
% Agreement on Schedule Code	97%	n.a.
Average Absolute AGI Difference	1,657	n.a.

Tableau A.5. Constrained Match, Improved Solution 1

FILE A RECORD	FILE B RECORD 1	2	3	4	
1	20 / 900	50 / 100	50	35	1000
2	20 / 500	100	10 / 1500	75	2000
3	85	5 / 300	115	30 / 200	500
	1400	400	1500	200	

Total solution distance = 55,500

WEIGHTED TABULATIONS

Description	This Merged File	Original Value
File A Record Data:		
Schedule code =1		3,000
Schedule code =2	Same	500
Total AGI (000s)		50,000
Total Deductions (000s)	Values	9,800
File B Record Data:		
Schedule code =1	As	2,900
Schedule code =2		600
Total AGI (000s)	Original	47,300
Transfer income (000s)		5,200
Avg. Family Size		2.65
Match Statistics (Wtd.)		
% Agreement on Schedule Code	97%	n.a.
Average Absolute AGI Difference	1,542	n.a.

Tableau A.6. Constrained Match, Improved Solution 2

FILE A RECORD	FILE B RECORD 1	2	3	4	
1	[20] 900	[50]	[50]	[35] 100	1000
2	[20] 500	[100]	[10] 1500	[75]	2000
3	[85]	[5] 400	[115]	[30] 100	500
	1400	400	1500	200	

Total solution distance = 51,500

WEIGHTED TABULATIONS

Description	This Merged File	Original Value
File A Record Data:		
Schedule code =1		3,000
Schedule code =2	Same	500
Total AGI (000s)		50,000
Total Deductions (000s)	Values	9,800
File B Record Data:		
Schedule code =1	As	2,900
Schedule code =2		600
Total AGI (000s)	Original	47,300
Transfer income (000s)		5,200
Avg. Family Size		2.65
Match Statistics (Wtd.)		
% Agreement on Schedule Code	97%	n.a.
Average Absolute AGI Difference	1,400	n.a.

Tableau A.7. Optimal Constrained Match

row and column totals, then the improved solution of Tableau A.5 could be obtained. Note that with this match, total distance drops by one-half, there is a 97 percent schedule code agreement and an improvement of $1285 in average AGI discrepancy, and the row and column totals are maintained. Further, by increasing w_{12} and dropping w_{22}, the next improved Tableau A.6 can be obtained, with a total distance lowered by 5000 and the average AGI discrepancy improved by $115 over that of Tableau A.5.

Finally, by increasing w_{14} and deleting w_{12}, the constrained-optimal solution of Tableau A.7 can be obtained. This match not only has a total solution value and schedule code agreement that are comparable to those of the unconstrained matches, but also has an average AGI discrepancy that is *better* than that of the unconstrained file-A match of Tableau A.3.

In summary, this small example illustrates that while unconstrained file matching can yield a closer overall match, serious data distortions can be introduced. These distortions can be overcome by constrained matching, which requires that the match-file record weights sum to the original individual record weights. And by devising a constrained-optimal match, not only are such data distortions eliminated, but extremely close matches at the record level can be obtained.

REFERENCES

Alter, Horst E. (1974). Creation of a Synthetic Data Set by Linking Records of the Canadian Survey of Consumer Finances with the Family Expenditure Survey 1970. *Annals of Economic and Social Measurement*, *(April)2*: 373–394.

Anderson, Erling B. (1980). *Discrete Statistical Models with Social Science Applications*, North-Holland, Amsterdam.

Armington, Catherine and Marjorie Odle (1975). Creating the MERGE-70 File: Data Folding and Linking. *Research on Microdata Files Based on Field Surveys and Tax Returns.* Working Paper I. The Brookings Institution (June). Mimeographed.

Barr, Richard S. (1981). Design of Experiments to Investigate Joint Distributions in Micro-Analytic Simulations, in *Proceedings of the 13th Conference on the Interface of Computer Science and Statistics*, Springer-Verlag, New York.

Barr, Richard S. (1982). Solution Strategies and Algorithm Behavior in Large-Scale Network Codes, in J. Mulvey, ed., *Evaluating Mathematical Programming Techniques*, Springer-Verlag, New York, 305–314.

Barr, Richard S., F. Glover, and D. Klingman (1977). The Alternating Basis Algorithm for Assignment Problems. *Mathematical Programming*, 13: 1–13.

Barr, Richard S., F. Glover, and D. Klingman (1978) A New Alternating Basis Algorithm for Semi-Assignment Networks, in W. W. White, ed., *Computers and Mathematical Programming*, U.S. Government Printing Office, Washington, D.C.

Barr, Richard S., F. Glover, and D. Klingman (1978). The Generalized Alternating Path Algorithm for Transportation Problems. *European J of Operational Research, 2*: 137–144.

Barr, Richard S., F. Glover, and D. Klingman (1979). Enhancements to Spanning Tree Labeling Procedures for Network Optimization. *INFOR, 17, 1*: 16–34.

Barr, Richard S. and J. Scott Turner (1978). A New Linear Programming Approach to Microdata File Merging, in *1978 Compendium of Tax Research*, sponsored by the Office of Tax Analysis, U.S. Department of the Treasury. (Barr and Turner's reply to Goldman also appears in that volume.)

Barr, Richard S. and J. Scott Turner (1978). *New Techniques for Statistical Merging of Microdata Files.* Paper prepared for the Conference on Microeconomics Simulation Models for the Analysis of Public Policy, National Academy of Sciences, March.

Barr, Richard S. and J. Turner (1980). New Techniques for Statistical Merging of Microdata Files, in R. Haveman and K. Hollenbeck, eds., *Microeconomic Simulation Models for the Analysis of Public Policy*, Academic Press, New York.

Barr, Richard S. and J. Scott Turner (1980). *Merging the 1977 Statistics of Income and the March 1978 Current Population Survey.* Prepared for the Office of Tax Analysis, U.S. Department of the Treasury.

Barr, Richard S. and J. Scott Turner (1981). Microdata File Merging Through Large-Scale Network Technology. *Mathematical Programming Studies, 5*: 1–22.

Budd, Edward C. (1971). The Creation of a Microdata File for Estimating the Size Distribution of Income. *Review of Income and Wealth*, (*December*) *17*: 317–333.

Budd, Edward C. (1972). Comments. *Annals of Economic and Social Measurement*, (*July*) *1*: 349–354.

Budd, Edward C. and Daniel B. Radner (1969). The OBE Size Distribution Series: Methods and Tentative Results for 1964. *American Economic Review*, (*May*) *59*: 435–449.

Budd, Edward C. and Daniel B. Radner (1975). The Bureau of Economic Analysis and Current Population Survey Size Distributions: Some Comparisons for 1964, in James D. Smith, ed., *The Personal Distribution of Income and Wealth, Studies in Income and Wealth, 39*: 449–558.

Budd, Edward C., Daniel B. Radner, and John C. Hinrichs (1973). *Size Distribution of Family Personal Income: Methodology and Estimates for 1964.* Bureau of Economic Analysis Staff Paper No. 21. U.S. Department of Commerce (June).

Colledge, M. J., J. H. Johnson, R. Pare, and I. G. Sande (1979). Large Scale Imputation of Survey Data. *1978 Proceedings of the American Statistical Association, Survey Research Methods Section*: 431–436.

Conover, W. J. (1971). *Practical Nonparametric Statistics*, John Wiley & Sons, New York.

Goldman, Alan J. (1978). Comment, in *1978 Compendium of Tax Research*, sponsored by the Office of Tax Analysis, U. S. Department of the Treasury.

Green, Paul E. (1978). *Analyzing Multivariate Data*, Dryden Press, Hinsdale, Illinois.

Greenberger, Martin, Matthew Crenson, and Brian Crissey (1976). *Models in the Policy Process*, Russell Sage Foundation, New York.

Kadane, Joseph B. (1975). *Statistical Problems of Merged Data Files.* OTA Paper 6, Office of Tax Analysis, U.S. Treasury Department (December 12).

Kadane, Joseph B. (1978). Some Statistical Problems in Merging Data Files, in *1978 Compendium of Tax Research*, sponsored by the Office of Tax Analysis, U.S. Department of the Treasury. (Kadane's reply to Sims also appears in that volume.)

Kilss, Beth and Fritz Scheuren (1978). *The 1973 CPS-IRS-SSA Exact Match Study: Past, Present, Future.* Paper presented at the NBER Workshop on the Uses of Social Security Research Files, March 15–17.

Makesh, S. and J. Scott Turner (1980). *Statistical Analysis of Oklahoma Microdata Files*, working paper, Office of Business and Economic Research, Oklahoma State University.

Minarik, Joseph J. (1980). The MERGE 1973 Data File, in R. H. Haveman and K. Hollenbeck, *Microeconomic Simulation Models for Public Policy Analysis*, Academic Press, New York.

Mulvey, John M. (1980). Reducing the U.S. Treasury's Taxpayer Data Base by Optimization. *Interfaces*, (*October*) *10*: 101–112.

Okner, Benjamin A. (1972). Constructing a New Data Base from Existing Microdata Sets: The 1966 Merge File. *Annals of Economic and Social Measurement*, (*July*) *1*: 325–342. (Okner's reply to comments also appear in that issue.)

Okner, Benjamin A. (1974). Data Matching and Merging: An Overview. *Annals of Economic and Social Measurement*, (*April*) *2*: 347–352.

Peck, Jon K. (1972). Comments. *Annals of Economic and Social Measurement*, (*July*) *1*: 347–348.

Radner, Daniel B. (1974). *The Statistical Matching of Microdata Sets: The Bureau of Economic Analysis 1964 Current Population Survey-Tax Model Match.* Ph.D. Dissertation, Department of Economics, Yale University. Microfilm.

Radner, Daniel B. (1978). *Age and Family Income.* Paper presented at the NBER Workshop on Policy Analysis with Social Security Research Files, Williamsburg, Virginia, March 15–17. Mimeographed.

Radner, Daniel B. (1978). The Development of Statistical Matching in Economics. *Proceedings of 1978 American Statistical Association, Social Science Section*, San Diego, August 16.

Radner, Daniel B. (1980). *An Example of the Use of Statistical Matching in the Estimation and Analysis of the Size Distribution of Income*, working paper, Office of Research and Statistics, Social Security Administration.

Radner, Daniel B. and Hans J. Muller (1978). Alternative Types of Record Matchings: Costs and Benefits. *1977 Proceedings of the ASA, Social Statistics Sections.*

Rao, C. Radhakrishna (1952). *Advanced Statistical Methods in Biometric Research*, John Wiley & Sons, New York.

Ruggles, Nancy and Richard Ruggles (1974). A Strategy for Merging and Matching Microdata Sets. *Annals of Economic and Social Measurement*, (*April*) *2*: 535–372.

Ruggles, Nancy, Richard Ruggles, and Edward Wolff (1977). Merging Microdata: Rationale, Practice and Testing. *Annals of Economic and Social Measurement*, (*Fall*) *6*: 429–444.

Siegel, Sidney (1956). *Nonparametric Statistics for the Behavioral Sciences*, McGraw-Hill, New York.

Sims, Christopher A. (1972). Comments. *Annals of Economic and Social Measurement*, (*July*) *1*: 343–346. (Sims' "Rejoinder" also appears in that issue.)

Sims, Christopher A. (1974). Comments. *Annals of Economic and Social Measurement*, (*April*) *2*: 395–398.

Sims, Christopher A. (1978). Comments on Kadane's Work on Matching to Create Synthetic Data, in *1978 Compendium of Tax Research*, sponsored by the Office of the Tax Analysis, U.S. Department of the Treasury.

Subcommittee on Matching Techniques. Federal Committee on Statistical Methodology (1980). *Report on Exact and Statistical Matching Techniques*, Statistical Policy Working Paper 5, U.S. Department of Commerce.

Turner, J. Scott and Gary E. Gilliam (1975). *Reducing and Merging Microdata Files*. OTA Paper 7, Office of Tax Analysis, U.S. Treasury Department (October).

Upton, Graham J. G. (1978). *The Analysis of Cross-Tabulated Data*, John Wiley & Sons, New York.

Wolff, Edward N. (1977). *Estimates of the 1969 Size Distribution of Household Wealth in the U.S. from a Synthetic Database*. Paper presented at the Conference on Research in Income and Wealth, Williamsburg, Virginia, December.

Yamane, Taro (1967). *Statistics, an Introductory Analysis*, Harper & Row, New York.

16

A Simulation Study of Selected Methods for Dealing with Survey Errors

GUNAR E. LIEPINS Energy Division, Oak Ridge National Laboratory, Oak Ridge, Tennessee

D. J. PACK School of Business Administration, Miami University, Oxford, Ohio

1. INTRODUCTION

Error localization is the inferential process of identifying the incorrect fields of inconsistent multivariate records. Error localization is required because questionnaire or survey data are often inconsistent (Pritzker et al., 1965; Williams, 1978; Juster, 1986; Andrews and Herzog, 1986). The Census Bureau (Gonzalez et al., 1975) has established standards for the presentation of inconsistent data, and Spencer (1985) has studied quality requirements for data as a function of their use. At times, error localization serves as a precursor to imputation (Madow et al., 1983; Wright, 1983; Little and Rubin, 1987; Rubin, 1987). More to the point, error localization in conjunction with quality-control methods, selective callbacks on specific records, and general quality consciousness can help to identify where the data problems seem to be most prevalent and may help to improve the quality of future data.

Error-localization methods fall into two classes: statistical methods and rule-based methods. For purposes of characterization, the proper domain of the statistical methods is the vast body of the data where little is known about individual records, whereas rule-based methods should be used to handle the individually important inconsistent records. Fellegi and Holt (1976) addressed statistical methods. They argued that errors should be attributed to the smallest possible number of components (hereafter called fields) of a multivariate record. While this objective is intuitively appealing, firmer grounding in empirical evidence and statistical theory seems warranted. Freund and Hartley (1967) and Naus, Johnson, and Montalvo (1972) previously had initiated work in this direction; however, neither addressed the minimum weighted fields to impute (MWFI) approach proposed by Fellegi and Holt. This paper evaluates three basic methods for error localization: MWFI, a discriminate function enhancement to MWFI (D-MWFI), and maximum posterior probability error localization (MPPEL).

The focus of this paper is twofold: to begin to bound the performance of the error-localization methods and to attempt to understand how selected data and error characteristics (such as the number of fields or the parameters of the error process) affect error-localization performance. For a narrowly chosen set of experiments, it is shown that MWFI is dominated by D-MWFI, which in turn is dominated by MPPEL. The characterization of localization performance as a function of data and error characteristics is far from complete. Preliminary observations suggest, for example, that the greater the error rate, the poorer the performance of error localization, and that the relative performance of MPPEL and the discriminate function is improved when the true data distribution is peaked rather than uniform. These and additional conclusions are documented in Section 5.

In an empirical study such as this, an important concern is the experimental design. Ideally, tests should be made with both real and simulated data. However, that ideal presupposes the availability of the correct counterparts to the inconsistent real data. The authors are unaware of such a resource, although Terry (1963) and Pritzker et al., (1965) have studied the frequency and patterns of errors. The fallback position is to use simulation exclusively, and that is what is done here.

An informal overview of the paper and the experimental design precedes the presentation of technical details. Data spaces and edits are chosen to represent consistency regions. "True" records are generated within those regions. These records are then perturbed until the resultant perturbations are no longer consistent. Next, the error-localization methods are applied to the inconsistent records. Since the true counterparts to the records are known, the performance of the methods can be analyzed.

It should be stressed that this chapter does not report comparative computation times of the various approaches, nor the technical details of how the approaches are actually implemented. In part, those results are available elsewhere, and the interested reader is referred to Liepins, Garfinkel, and Kunnathur (1982); Garfinkel, Kunnathur, and Liepins (1986); and Garfinkel, Kunnathur, and Liepins (1988).

2. DATA AND ERROR MODELS

For purposes of this chapter, a record y is a questionnaire response. Thus, if there are n questions, y is an n-dimensional vector and the jth field refers to the jth question. Generally, records (data) can be continuous, discrete, or mixed. In this chapter, attention will be restricted to discrete data.

The model of the data process specifies that the observed record $\mathbf{y}$ is the true value $\mathbf{x}$ plus an error $\boldsymbol{\varepsilon}$,

$$\mathbf{y} = \mathbf{x} + \boldsymbol{\varepsilon} \tag{1}$$

The multivariate data space $D = \Pi D_i$ (the Cartesian product of the univariately allowable entries) is the set of all potentially observable records $\mathbf{y}$ and, for purposes of this study, is limited to a finite space. Associated with the data space D and the data process are a set of consistency constraints $C = \{c_i\}$. These constraints partition the data space into complementary regions: the consistency region D_C and its complement $\tilde{D}_C$. Thus, a record $\mathbf{y} \in \tilde{D}_C$ is known to be inconsistent and, therefore, incorrect; whereas a record $\mathbf{y} \in D_C$ is consistent but not necessarily correct. The error-localization techniques studied here are classified as deterministic insofar as they only address inconsistent data; oulier methods are not considered. Associated with a given data process is a (unknown in reality) density function $p(\mathbf{x})$ of true values $\mathbf{x} \in D_C$. Associated with the multivariate error process $\boldsymbol{\varepsilon}$ is a family of functions $\{p(\mathbf{y}|\mathbf{x})\}$ that specify the conditional probabilities that records $\mathbf{y}$ are observed when the true record is $\mathbf{x}$. In this chapter, the simplifying assumption is made that the density functions $p(\mathbf{y}|\mathbf{x})$ can be characterized by the probability of error p_i in each of the fields i (possibly conditional on some other field j) and the conditional distribution of the magnitude of the error, given that an error has been made. These parameters, the data space D, the consistency region D_C, and the density functions $p(\mathbf{x})$ and $p(\mathbf{y}|\mathbf{x})$ are varied in the simulations reported here.

EXAMPLE 1 Consider a two-item questionnaire. (In this admittedly artificial example, college XYZ is a prestigious college. Many individuals, former students or not, would like to join the alumni association, but only graduates of the college are accepted.)

Question 1 Did you attend college XYZ?

a. Y b. N

Question 2 Was your application to the XYZ alumni association accepted?

a. Y b. N c. N/A

The data space can be represented by a 2×3 matrix, as shown in Table 1. The data space is $D = D_1 \times D_2 = \{Y, N\} \times \{Y, N, N/A\}$. The consistency region D_C consists of the squares labeled YY, YN, YN/A, NN, NN/A. If all consistent responses are equally likely, then the distribution $p(\mathbf{x})$ of the true values over the consistency region D_C is uniform. Assume that the probability p_1 of answering the first question incorrectly is 0.25 and that the conditional probability $(p_2|q_1)$ of answering the second question incorrectly given that the first was answered correctly is 0.1. Otherwise, assume the conditional probability $(p_2|\bar{q}_1) = 0.6$. If the second question is answered incorrectly, then assume that all possible wrong answers for the second question are equally likely. With these assumptions, it is not hard to verify that the distribution $p(\mathbf{x})$ of true records $\mathbf{x}$ and the expected distribution $p(\mathbf{y}) = \Sigma_{\mathbf{x}} p(\mathbf{y}|\mathbf{x})p(\mathbf{x})$ of observed records y are given by Tables 2 and 3, respectively.

Table 1. Data Space for Questionnaire

		Accepted		
		Y	N	N/A
Attended College	Y	YY	YN	YN/A
	N	NY	NN	NN/A

Table 2. Probability Distribution $p(\mathbf{x})$ of True Values $\mathbf{x}$

		Accepted		
		Y	N	N/A
Attended College	Y	0.2	0.2	0.2
	N	0	0.2	0.2

Table 3. Expected Probability Distribution of Observed Values $\mathbf{y}$

		Y	N	N/A
Attended College	Y	.18	.185	.185
	N	.065	.1925	.1925

3. ERROR-LOCALIZATION METHODS

3.1 Minimum Weighted Fields to Impute (MWFI)

Let $\mathbf{y}$ be an observed inconsistent record. The Fellegi-Holt (Fellegi and Holt 1976; Sande 1976) formulation of MWFI is to perturb the minimum number of fields to render the record $\mathbf{y}$ consistent.

Set $w_i = 1$ for all $i = 1, \ldots, n$ and $\boldsymbol{\varepsilon} = (\varepsilon_1, \ldots, \varepsilon_n)$

$$\text{Minimize (over } \boldsymbol{\varepsilon}) \sum_{i=1}^{n} w\sigma(\varepsilon_i) \tag{2}$$

$$\text{subject to } \mathbf{y} + \boldsymbol{\varepsilon} \text{ is consistent, and} \tag{3}$$

$$\sigma(\varepsilon_i) = \begin{cases} 0 & \text{if and only if } \varepsilon_i = 0 \\ 1 & \text{otherwise} \end{cases} \tag{4}$$

The weights w_i in (2) generally need not be chosen equal, and intuitively they should be inversely proportional to the probability of error in the respective fields. In any case, the weights must be chosen by the analyst performing the error localization. This chapter investigates MWFI with all weights w_i equal to one, with weights defined as $w_i = \log(1 - p_i) - \log(p_i)$, where p_i is the prior probability of error in field i, and with weights defined as the inverses of those associated with the prior probabilities.

3.2 Maximal Prior Probability Error Localization (MAXP)

For an inconsistent record **y**, the maximal prior probability error localization (MAXP) provides one conceptual way to infer the incorrect fields. MAXP identifies a set s of fields that maximizes the prior probability that changes in the fields specified by s alone can render the record consistent. Because prior error probabilities are generally not known, MAXP is not considered a practical error-localization technique. It serves merely to suggest the plausibility of MWFI. Assume that the probability of an error in field i is p_i and that errors in fields are independent. Then, the probability that exactly the fields indexed by s are wrong is given by $\Pi_{i \in s} p_i \Pi_{i \notin s}(1 - p_i)$. The negative log transformation of a normalized factor of the prior probability $\Pi_{i \in s} p_i \Pi_{i \notin s}(1 - p_i)$ suggests weights $w_i = \log(1 - p_i) - \log(p_i)$ for MWFI, as specified in (2).

3.3 Maximal Posterior Probability Error Localization (MPPEL)

MPPEL is a Bayes-rule error localization. As with MAXP, MPPEL is not considered a practical error-localization method. (MPPEL generally is computationally impractical, even if the required probabilistic information were available.) Its usefulness to this study is that it is the theoretically optimal record-by-record optimization method (in terms of one of the performance characteristics considered) and serves as a standard against which the other methods can be compared.

Consider an observed inconsistent, n-dimensional record **y** and let S represent the power set of the indices $1, \ldots, n$. Then, MPPEL determines the incorrect fields as those indexed by s°, where s° maximizes the expression.

$$\underset{s \in S, x \in D_C}{\text{maximize}} \left\{ \sum p(\mathbf{x}|\mathbf{y}) : x_i = y_i \text{ for } i \notin s,\ x_i \neq y_i \text{ for } i \in s \right\} \tag{5}$$

That is, changes are made only to those fields indexed by s.

For any inconsistent record **y** and the error models considered in this chapter, the error localization specified by MPPEL is at least as likely as any other error localization. This result is straightforward: Note that for the processes considered, $f(\mathbf{x}, \boldsymbol{\varepsilon}) = \mathbf{x} + \boldsymbol{\varepsilon}$ are one-to-one on each of the components separately. That is, each of $f_x(\boldsymbol{\varepsilon}) = f(\mathbf{x}, \cdot)$ and $f_\varepsilon(\ ) = f(\cdot, \boldsymbol{\varepsilon})$ is one-to-one. Let $s \in S$ and $p(s, \mathbf{y})$ be the probability that s is the true localization for the observation **y**. Then, $p(s|\mathbf{y}) = \Sigma_s p(\boldsymbol{\varepsilon}, \mathbf{y})$, where the latter sum is taken over all error patterns specified by s. Since $f(\mathbf{x}, \boldsymbol{\varepsilon})$ is one-to-one on each variable separately, $\Sigma_s p(\boldsymbol{\varepsilon}, \mathbf{y}) = \Sigma_{\mathbf{x} \in D_c} \{p(\mathbf{x}|\mathbf{y}) : x_i = y_i \text{ for } i \notin s, x_i \neq y_i \text{ for } i \in s\}$.

Theoretically, given (generally unavailable) distributional information about the data process,

$$\mathbf{y} = \mathbf{x} + \boldsymbol{\varepsilon}$$

in the form of the probabilities

$$p(\mathbf{y}|\mathbf{x}) \text{ and } p(\mathbf{x})$$

MPPEL would be implemented through the use of Bayes theorem

$$p(\mathbf{x}|\mathbf{y}) = \frac{p(\mathbf{y}|\mathbf{x})\,p(\mathbf{x})}{p(\mathbf{y})} \tag{6}$$

Since the required maximization is being done on a record-by-record basis, the denominator can be ignored. For this study, $p(\mathbf{x})$ and $p(\mathbf{y}|\mathbf{x})$ are determined according to the theoretically specified (known within the simulation but unknown in practice) data and error models.

EXAMPLE 2 Consider the data set and distributions of Example 1 and an observed record NY. This record is known to be inconsistent and, by Bayes' theorem, the conditional probabilities of Table 4 can be calculated. A change in just the first field corresponds to the "correction" NY→YY with Bayes probability of 0.31. The MPPEL value for the second field alone is 0.23 (=0.115 + 0.115) and for both components simultaneously is 0.46 (=0.23 + 0.23). Thus, the MPPEL error localization identifies both fields as incorrect.

Table 4. Probabilities of True Values Given Observation NY

Value = $\mathbf{x}$	Probability = $p(\mathbf{x}|\mathbf{y} = \text{NY})$
YY	0.31
YN	0.23
YN/A	0.23
NY	0
NN	0.115
NN/A	0.115

3.4 Discriminate Function Enhanced MWFI (D-MWFI)

One of the properties of MWFI (encountered if all weights w_i are equal) is that multiple alternative solutions are often found. One way to deal with multiple solutions is to choose one at random. Another is to perturb the weights slightly so they are no longer equal. Neither uses the "information in the data." A discriminate function is hoped to lead to better performance than does random choice. Such a discriminate function is first motivated and its implementation is then described.

Consider some simplifying assumptions about the data process $\mathbf{y} = \mathbf{x} + \boldsymbol{\varepsilon}$:

$$p\{\varepsilon_i \neq 0 | \varepsilon_j \neq 0\} = p\{\varepsilon_i \neq 0\} = p_i \text{ for } i \neq j \tag{7}$$

For each $i = 1, \ldots, n$, ε_i has uniform distribution for all i over the set of nonzero feasible values for ε_i. (8)

Let $w(s, y)$ be the set of possible consistent records whose ith component equals y_i for $i \in s$; that is, $w(s, \mathbf{y}) = \{\mathbf{x} : \mathbf{x} \in D_C,\ x_i = y_i \text{ for } i \notin s$ and $x_i \neq y_i$ for $i \in s\}$. Define $|w(s, \mathbf{y})|$ to be the empirical probability measure of the set $w(s, y)$; that is, $|w(s, y)| = \Sigma p(\mathbf{x})$, where the sum is taken over all $\mathbf{x}$ in $w(s, \mathbf{y})$.

Let $|D_i|$ be the cardinality of D_i (without loss of generality, $|D_i| \geqslant 2$). Then, for any inconsistent record $\mathbf{y}$ (under the assumptions of independence of errors and uniform magnitude of errors (7) and (8), respectively), MPPEL—i.e., maximization of (5) is equivalent to maximization of

$$\frac{|w(s, \mathbf{y})|}{\Pi_{i \in s}(|D_i| - 1)} \left[\prod_{i \in s} p_i \prod_{i \notin s} (1 - p_i) \right] \tag{9}$$

For weights associated to MAXP, the term within the brackets in expression (9) reduces to MWFI under the negative log transformation. Therefore, it seems reasonable to distinguish between alternative optimal MWFI solutions by the discriminate function

$$\frac{|w(s, \mathbf{y})|}{\Pi_{i \in s}(|D_i| - 1)} \tag{10}$$

In practice, the numerator $|w(s, \mathbf{y})|$ of (10) is unknown and must be estimated by sampling. The tacit assumption must be made that the distribution of observed records $\mathbf{y}$ provides an adequate proxy for the distribution of true records. For the purposes of this chapter, the complete population of observed records $\mathbf{y} \in D_C$ was used to estimate $|w(s, \mathbf{y})|$.

4. SIMULATION

A systematic simulation was undertaken to study five error-localization methods:

1. MWFI with $w = 1$, (1-MWFI)
2. MWFI with $w = 1$ and ties (partially) resolved by the discriminate function (10), (D-MWFI)
3. MWFI with weights chosen according to the negative log transformation of the prior probabilities (P-MWFI)
4. MWFI with weights chosen to be the reciprocal of those specified by P-MWFI (RP-MWFI)
5. MPPEL

It was anticipated that 1-MWFI performance would prove to be relatively insensitive to error model, but would be dominated by both P-MWFI and D-MWFI. Each of these was anticipated to be dominated by MPPEL. RP-MWFI was investigated to determine the effect of weight misspecification; it was anticipated to be the worst of the methods.

The underlying experimental design consisted of two data sets—the first with six fields and the second with four. The full technical description of the experimental design is provided in Appendix 1 and is summarized in the paragraphs below.

4.1 First Data Set Experimental Designs

The first data set experimental designs incorporated four additional factors: edit set, distribution of true data, error probability, and distribution of error.

The three edit sets for the first data set specified 433, 245, and 212 of the 576 data points as inconsistent. For each of these edit sets, uniform, unimodal, and multimodal distributions of true data were investigated with the anticipation that the relative effectiveness of the discriminate function would decrease as modality varied from unimodal to multimodal and further to uniform.

Error probability was determined according to one of six models, ".95," ".9," ".8," ".6/.95," mild dependency (DEMIL), and strong dependency (DESTR). Models ".95," ".9," and ".8" specify that the prior probability of no error in each field would be .95, .9, and .8, respectively, and that the incidence of error in any one field would be independent of the other fields. Formally, for ".95," this is the model described by $p\{\varepsilon_i \neq 0 | \varepsilon_j \neq 0\} = p\{\varepsilon_i \neq 0\} = p_i = 0.05$. The other models are:

Model .6/.95 specifies $p_1 = p_2 = 0.4$ and $p_i = 0.05$ otherwise.
Model DEMIL (mild dependency) specifies $p_i = 0.05$ for $i = 1, 3, 5, 6$, $p(\varepsilon_2 \neq 0|\varepsilon_1 \neq 0) = 0.4 = p(\varepsilon_4 \neq 0|\varepsilon_3 \neq 0)$, and $p(\varepsilon_2 \neq 0|\varepsilon_1 = 0) = 0.05 = p(\varepsilon_4 \neq 0|e_3 = 0)$.
Model DESTR (strong dependency) specifies $p_i = 0.05$ for $i = 1, 3, 5$, $p(\varepsilon_2 \neq 0|\varepsilon_1 \neq 0) = p(\varepsilon_4 \neq 0|\varepsilon_3 \neq 0) = p(\varepsilon_6 \neq 0|\varepsilon_5 \neq 0) = 0.8$, and $p(\varepsilon_2 \neq 0|\varepsilon_1 = 0) = p(\varepsilon_4 \neq 0|\varepsilon_3 = 0) = p(\varepsilon_6 = 0|\varepsilon_5 = 0) = 0.05$.

Error distribution was specified according to one of two models: uniform error or nonuniform error. For the uniform error (conditioned on the existence of an error in field i), the error ε_i was chosen to have uniform distribution over the set of nonzero feasible values for ε_i. The nonuniform distribution was specified by the probability distribution where $\varepsilon_i = \tau$ is chosen with probability inversely proportional to τ^2.

4.2 Second Data Set Experimental Designs

The second data set had four fields and was simulated with only one edit set specifying 749 of the possible 1044 data points as inconsistent. The remaining factors were varied as in the first data set. Similar to the first data set, the distributions of true data for the second data set were varied over uniform, unimodal, and multimodal. The error models ".95," ".9," and ".8" were the obvious counterparts to the respective models for the first data set. The other models were:

Model .6/.95 was specified as $p_1 = p_2 = 0.4$ and $p_i = 0.05$ otherwise.
Model DEMIL was specified as $p_i = 0.05$ for $i = 1, 3, 4$, $p(\varepsilon_2 \neq 0|\varepsilon_1 \neq 0) = 0.4$, and $p(\varepsilon_2 \neq 0|\varepsilon_1 = 0) = 0.05$.

Model DESTR was specified as $p_i = 0.05$ for $i = 1$ and 3, $p(\varepsilon_2 \neq 0|\varepsilon_1 \neq 0) = p(\varepsilon_4 \neq 0|\varepsilon_3 \neq 0) = 0.8$, and $p(\varepsilon_2 \neq 0|\varepsilon_1 = 0) = p(\varepsilon_4 \neq 0|\varepsilon_3 = 0) = 0.05$.

The error distributions were the counterparts to those tested for the first data set: uniform error and nonuniform error.

4.3 Experimental Procedure

Figure 1 gives a flowchart of the procedure executed for each of the test cases studied. For each case, the simulation generated 10,000 inconsistent records and evaluated the various error-localization methods. The record-

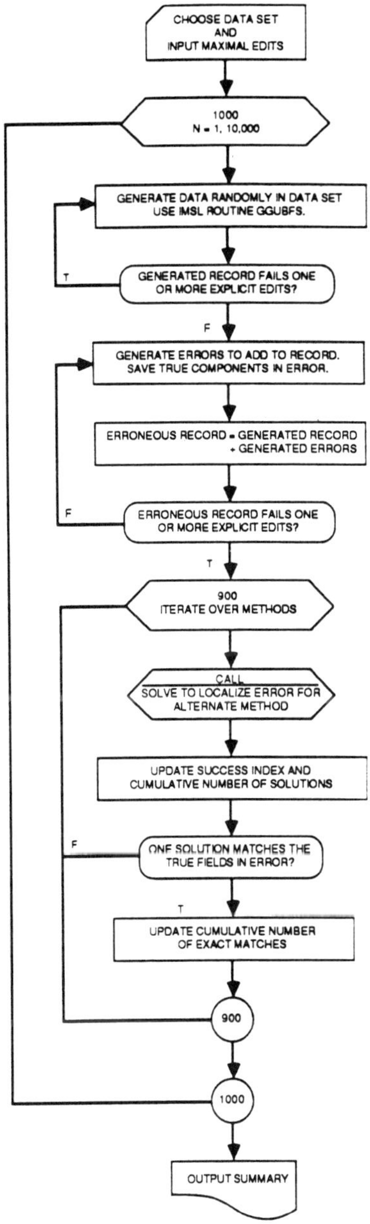

Figure 1. Flowchart of simulation evaluation of error-localization techniques.

by-record procedure was:

1. Generate a record **x** from the assumed distribution of true values. If the record **x** is consistent, go to 2. Otherwise repeat 1.
2. Perturb the record **x** by the introduction of errors generated from the assumed error model. Define the perturbed record to be **y**.
3. If the perturbed record **y** fails one or more explicit edits, proceed. If not, reperturb—that is, return to step 2 with the original record **x**.
4. Localize the error according to each of the methods to be studied.

Five methods for error localization (1-MWFI, D-MWFI, P-MWFI, RP-MWFI, and MPPEL) were systematically studied, and a wide variety of other methods were tentatively investigated.

4.4 Evaluation Criteria

The performance of the error-localization methods is given in terms of criteria defined in this section:

Solution—The set(s) of fields specified by a localization method. The total number of solutions for a test case and given method will be denoted t; for any test case and any method, $t \geqslant 10{,}000$. Often, more than one solution will be found for a given inconsistent record. Therefore, t can be substantially greater than 10,000.

Match—A solution containing all the fields known to be in error and no other fields.

Matches/Solution (M/S)—The number of matches divided by the number of solutions.

Success Index (PFCD, Percentage of Fields Correctly Dealt with)—The simple average over all solutions of the proportion of fields correctly dealt with. For each solution, the contribution to the numerator is determined as (number of fields suggested to be in error that actually were) + (number of fields suggested not to be in error that actually were not). The overall denominator is nt, the product of the dimension of the space times the number of solutions.

(The matches/solution statistic and success index statistic are suggested to be the primary measures for comparing the different localization methods.)

True Proportion of Correct Fields (tp)—Proportion of correct fields averaged over all 10,000 inconsistent records **y**. The true proportion of correct fields is thought to be a major determinant of the performance of the various error-localization methods: A minimal field's approach is intuitively plausible when errors have low probability and are independent; otherwise, the approach seems less plausible.

EXAMPLE 3 Consider the first data space with the first explicit edit set.

1. Assume that the generated record is $\mathbf{x} = (0, 1, 1, 0, 1, 2)$. Check $\mathbf{x} \notin e_i$, $i = 1, \ldots, 5$. Go to step 2 and perturb.
2. Assume that the first induced perturbation is $(0, 2, 0, 0, 0, 0)$. Then, the perturbed record is $\mathbf{y} = [0, 3(\text{mod}3), 1, 0, 1, 2] = (0, 0, 1, 0, 1, 2)$. Check $\mathbf{y} \notin e_i$, $i = 1, \ldots, 5$. Reperturb $\mathbf{x}$.
3. Assume that the next induced perturbation is $(1, 2, 1, 0, 0, 2)$. Then, the perturbed record is $\mathbf{y} = [1, 3(\text{mod}3), 2(\text{mod}2), 0, 1, 4(\text{mod}4)] = (1, 0, 0, 0, 1, 0)$. Check $\mathbf{y} \in e_2$ and $\mathbf{y} \in e_3$.

(Errors were introduced into components 1, 2, 3, and 6, and edits e_2 and e_3 are failed.)

4. For the case of 1-MWFI, there are nine solutions with the following components:
   a. 123
   b. 125
   c. 126
   d. 136
   e. 156
   f. 234
   g. 235
   h. 346
   i. 456

There are no matches. The success index (PFCD) is computed as follows:

a. 5/6
b. 3/6
c. 5/6
d. 5/6
e. 3/6
f. 3/6
g. 3/6
h. 3/6
i. 1/6

---

avg. 31/54

Cumulatively, 31 of the 54 fields were correctly dealt with over the nine different solutions.

4.5 Notation

For the presentation of the findings, the following notation will be used:

```
Dm En UF [xxxxxx] U
      UM         N
      MM
```

where

m = example number (1, 2)
n = edit set number (1, 2, 3 Ex. 1; 1 Ex. 2)
UF = uniform
UM = unimodal
MM = multimodal
xxxxxx = .95, .9, .8, .6/.95, DEMIL, DESTR
U = uniform error
N = nonuniform error

For example, data set 1, edit set 2, unimodal true data, error model .6/.95, nonuniform error would be denoted D1E2UM[.6/.95]N.

5. FINDINGS

Results and conclusions from this study fall into three categories: (1) results and general hypotheses with supporting evidence, (2) conjectures, and (3) empirical observations without ready explanations.

5.1 Results and Hypotheses

The general hypotheses and conclusions suggested are single factored. Little is known regarding combinations of factors.

1. The localization methods 1-MWFI, P-MWFI, and RP-MWFI all can be proven to yield the same error localizations for the independent error models .95, .9, and .8.
2. For a given data/edit-set combination and data model, the performance of all localization methods deteriorates as the error rate increases. Similarly, the incremental improvement of D-MWFI over 1-MWFI deteriorates with increasing error rate. This hypothesis is borne out by the empirical data that suggest that performance can be roughly ranked in decreasing order of the error models .95, .9, DEMIL, .6/.95, DESTR, and .8 (see Tables 5 and 6).

Table 5. M/S Performance as a Function of Error Model (Averaged Over True Distributions and Uniform and Nonuniform Error)

	.95	.9	DEMIL	.6/.95	DESTR	.8
D1E1						
tp	.800	.765	.767	.737	.712	.693
1-MWFI	.340 (.31–.36)	.272 (.24–.3)	.270 (.24–.29)	.198 (.17–.23)	.180 (.15–.20)	.172 (.15–.19)
D-MWFI	.620 (.48–.75)	.502 (.39–.62)	.482 (.36–.61)	.337 (.26–.46)	.297 (.22–.40)	.317 (.25–.40)
MPPEL	.625 (.50–.75)	.502 (.39–.62)	.537 (.41–.64)	.528 (.49–.58)	.573 (.38–.62)	.330 (.28–.41)
D1E2						
tp	.790	.755	.763	.708	.710	.683
1-MWFI	.286 (.26–.30)	.230 (.20–.24)	.245 (.23–.25)	.145 (.13–.16)	.168 (.16–.19)	.140 (.12–.15)
D-MWFI	.512 (.42–.66)	.398 (.33–.53)	.463 (.40–.58)	.182 (.12–.24)	.330 (.25–.43)	.233 (.19–.32)
MPPEL	.477 (.36–.66)	.377 (.29–.53)	.463 (.37–.55)	.233 (.21–.29)	.452 (.33–.58)	.235 (.19–.33)
D1E3						
tp	.798	.765	.770	.705	.713	.695
1-MWFI	.215 (.21–.22)	.175 (.17–.18)	.185 (.18–.19)	.113 (.11–.12)	.113 (.10–.12)	.118 (.11–.13)
D-MWFI	.465 (.31–.65)	.367 (.23–.54)	.395 (.26–.55)	.143 (.11–.20)	.245 (.11–.38)	.240 (.16–.36)
MPPEL	.443 (.31–.65)	.377 (.23–.54)	.415 (.25–.56)	.270 (.20–.37)	.395 (.24–.51)	.235 (.15–.36)
D2E1						
tp	.718	.685	.688	.618	.573	.620
1-MWFI	.517 (.49–.54)	.453 (.42–.47)	.463 (.45–.48)	.367 (.33–.41)	.232 (.17–.30)	.342 (.31–.37)
D-MWFI	.783 (.71–.83)	.688 (.61–.74)	.497 (.47–.51)	.412 (.35–.48)	.283 (.18–.38)	.512 (.44–.56)
MPPEL	.792 (.70–.83)	.702 (.63–.76)	.730 (.65–.78)	.575 (.49–.63)	.640 (.55–.70)	.537 (.45–.59)

Table 6. PFCD Performance as a Function of Error Model (Averaged Over True Distributions and Uniform and Nonuniform Error)

	.95	.9	DEMIL	.6/.95	DESTR	.8
D1E1						
tp	.8	.765	.767	.737	.712	.693
1-MWFI	.783 (.76–.80)	.763 (.75–.77)	.758 (.75–.77)	.737 (.73–.74)	.720 (.71–.73)	.703 (.69–.71)
D-MWFI	.884 (.83–.93)	.853 (.80–.90)	.843 (.79–.90)	.807 (.75–.87)	.793 (.74–.80)	.782 (.74–.89)
MPPEL	.883 (.84–.93)	.858 (.81–.91)	.860 (.80–.91)	.885 (.86–.91)	.845 (.78–.90)	.795 (.75–.85)
D1E2						
tp	.790	.755	.763	.708	.710	.683
1-MWFI	.768 (.76–.77)	.738 (.73–.74)	.747 (.74–.75)	.708 (.70–.71)	.697 (.69–.71)	.682 (.67–.69)
D-MWFI	.853 (.82–.91)	.827 (.79–.87)	.825 (.79–.88)	.730 (.67–.77)	.780 (.74–.84)	.745 (.72–.80)
MPPEL	.842 (.80–.91)	.810 (.77–.88)	.823 (.78–.89)	.765 (.75–.79)	.797 (.74–.87)	.745 (.71–.81)
D1E3						
tp	.798	.765	.770	.705	.713	.695
1-MWFI	.730 (73)	.707 (.70–.71)	.717 (.71–.72)	.670 (.66–.68)	.667 (.66–.67)	.655 (.65–.66)
D-MWFI	.830 (.77–.90)	.800 (.74–.87)	.805 (.74–.87)	.688 (.66–.72)	.752 (.68–.83)	.742 (.69–.81)
MPPEL	.825 (.75–.90)	.798 (.73–.87)	.803 (.73–.87)	.780 (.75–.83)	.770 (.65–.85)	.743 (.68–.81)
D2E1						
tp	.718	.685	.688	.618	.573	.620
1-MWFI	.797 (.78–.81)	.767 (.75–.78)	.765 (.76–.79)	.767 (.74–.79)	.712 (.71–.72)	.712 (.70–.73)
D-MWFI	.912 (.87–.93)	.887 (.85–.91)	.892 (.85–.91)	.841 (.83–.86)	.792 (.75–.80)	.817 (.78–.84)
MPPEL	.917 (.86–.94)	.890 (.84–.91)	.912 (.85–.93)	.868 (.83–.89)	.872 (.81–.90)	.835 (.79–.87)

3. The more "peaked" the distribution of the true data, the greater the performance increment between D-MWFI and 1-MWFI. This hypothesis is supported by the empirical data that suggest that the increment in performance is greatest for the unimodal cases, somewhat less for the multimodal cases, and least for the uniform cases (see Table 7).
4. The "better tuned" the weights in the MWFI error-localization method, the better the performance. The danger of tuning the weights is that misspecification can dramatically degrade performance. The simulation results demonstrate that for error model .6/.95, P-MWFI dominates 1-MWFI, which in turn dominates RP-MWFI (the only error model where the dominance can be strictly expected theoretically), except in cases D1E2UM[.6/.95]U and D1E2UM[.6/.95]N. In these latter two cases, the dominance is inexplicably reversed. As expected, the misspecified localization RP-MWFI performs particularly poorly for the error model .6/.95 (see representative Tables 8 and 9 for M/S and PFCD performance for D1E1.
5. For the M/S performance criterion, MPPEL is the (unrealizable) optimal localization method. For those cases in which the P-MWFI or D-MWFI performance approaches MPPEL, these methods provide realizable near-optimal localization. This occurs for error models .95, .9, and .8 of the simulation.

5.2 Conjectures

The performance of all localization methods is thought to deteriorate as the number of fields increases. Moreover, the incremental improvement due to the discriminate function is also thought to decrease. No formal proof of this conjecture is provided, nor do the simulations address this issue.

5.3 Empirical Observations

1. D-MWFI performs comparably to MPPEL on the independent uniform error models .95, .9, and .8. On the other hand, the relative performance of MPPEL is substantially superior on DEMIL and DESTR (see Table 10).
2. The performance of error localization as a function of data/edit sets can be ordered generally as D2E1, D1E1, D1E2, and D1E3. This order can be verified by the M/S performance and PFCD performance tables (Tables 5 and 6) on a case-by-case basis.

Table 7. Average Increment in Performance of D-MWFI Over 1-MWFI (Averaged Over Six Error Models)

	M/S	PFCD
D1E1		
UMU	.300	.140
UMN	.270	.132
BMU	.202	.085
BMN	.178	.083
UFU	.095	.032
UFN	.077	.027
D1E2		
UMU	.230	.118
UMN	.200	.098
BMU	.153	.065
BMN	.122	.052
UFU	.087	.048
UFN	.090	.037
D1E3		
UMU	.277	.133
UMN	.243	.112
BMU	.128	.077
BMN	.182	.098
UFU	.045	.033
UFN	.062	.045
D2E1		
UMU	.218	.117
UMN	.220	.128
BMU	.210	.107
BMN	.220	.117
UFU	.132	.057
UFN	.145	.067

Table 8. M/S Performance for Data Set 1, Edit Set 1

1-MWFI	D-MWFI	MPPEL	1-MWFI	D-MWFI	MPPEL	1-MWFI	D-MWFI	MPPEL
0.36	*UF[0.95]U* tp : 0.8 0.51	0.50	0.30	*UF[0.9]U* tp : 0.77 0.42	0.42	0.19	*UF[0.8]U* tp : 0.7 0.28	0.28
0.35	*UF[0.95]N* tp : 0.8 0.48	0.51	0.28	*UF[0.9]N* tp : 0.76 0.39	0.42	0.18	*UF[0.8]N* tp : 0.69 0.25	0.28
0.34	*UM[0.95]U* tp : 0.80 0.75	0.75	0.27	*UM[0.9]U* tp : 0.77 0.62	0.62	0.17	*UM[0.8]U* tp : 0.70 0.40	0.41
0.31	*UM[0.95]N* tp : 0.80 0.71	0.71	0.24	*UM[0.9]N* tp : 0.76 0.56	0.57	0.15	*UM[0.8]N* tp : 0.68 0.34	0.35
0.35	*MM[0.95]U* tp : 0.80 0.65	0.64	0.28	*MM[0.9]U* tp : 0.77 0.53	0.52	0.18	*MM[0.8]U* tp : 0.70 0.34	0.34
0.33	*MM[0.95]N* tp : 0.80 0.62	0.64	0.26	*MM[0.9]N* tp : 0.76 0.49	0.51	0.16	*MM[0.8]N* tp : 0.69 0.29	0.32

Table 8. (Continued)

I-MWFI	D-MWFI	P-MWFI	RP-MWFI	MPPEL	1-MWFI	D-MWFI	P-MWFI	RP-MWFI	MPPEL
0.23	0.27	*UF[0.6/0.95]U* tp:0.75 0.41	0.05	0.49	0.24	0.54	*UM[DEMIL]N* tp:0.76 0.14	0.37	0.62
0.22	0.26	*UF[0.6/0.95]N* tp:0.74 0.40	0.04	0.50	0.28	0.52	*MM[DEMIL]U* tp:0.77 0.24	0.35	0.55
0.19	0.46	*UM[0.6/0.95]U* tp:0.74 0.29	0.06	0.54	0.25	0.46	*MM[DEMIL]N* tp:0.76 0.19	0.33	0.57
0.17	0.42	*UM[6.0/0.95]N* tp:0.73 0.27	0.04	0.58	0.20	0.26	*UF[DESTR]U* tp:0.72 0.25	0.20	0.33
0.20	0.32	*MM[6.0/0.95]U* tp:0.73 0.31	0.06	0.49	0.19	0.22	*UF[DESIR]U* tp:0.71 0.22	0.19	0.45
0.18	0.29	*MM[0.6/0.95]N* tp:0.73 0.30	0.04	0.57	0.19	0.40	*UM[DESIR]U* tp:0.72 0.24	0.18	0.62
0.29	0.40	*UF[DEMIL]U* tp:0.77 0.29	0.34	0.41	0.15	0.31	*UM[DESTR]N* tp:0.70 0.17	0.17	0.61
0.28	0.36	*UF[DEMIL]N* tp:0.77 0.27	0.33	0.43	0.19	0.33	*MM[DESTR]U* tp:0.72 0.23	0.19	0.54
0.28	0.61	*UM[DEMIL]U* tp:0.77 0.21	0.37	0.64	0.16	0.26	*MM[DESTR]N* tp:0.70 0.18	0.18	0.60

Table 9. PFCD Performance for Data Set 1, Edit Set 1

1-MWFI	D-MWFI	MPPEL	1-MWFI	D-MWFI	MPPEL	1-MWFI	D-MWFI	MPPEL
0.36	*UF[0.95]U* tp : 0.8 0.51	0.50	0.30	*UF[0.9]U* tp : 0.77 0.42	0.42	0.19	*UF[0.8]U* tp : 0.7 0.28	0.28
0.80	*UF[0.95]U* tp : 0.8 0.84	0.84	0.77	*UF[0.9]U* tp : 0.77 0.81	0.81	0.71	*UF[0.8]U* tp : 0.7 0.75	0.75
0.79	*UF[0.95]N* tp : 0.8 0.83	0.85	0.77	*UF[0.9]N* tp : 0.76 0.80	0.82	0.71	*UF[0.8]N* tp : 0.69 0.74	0.75
0.78	*UF[0.95]U* tp : 0.80 0.93	0.93	0.76	*UM[0.9]U* tp : 0.77 0.90	0.91	0.70	*UF[0.8]U* tp : 0.70 0.84	0.85
0.77	*UM[0.95]N* tp : 0.80 0.92	0.92	0.75	*UM[0.9]N* tp : 0.76 0.85	0.85	0.69	*UM[0.8]N* tp : 0.68 0.82	0.82
0.79	*MM[0.95]U* tp : 0.80 0.89	0.89	0.77	*MM[0.9]U* tp : 0.77 0.86	0.86	0.71	*MM[0.8]U* tp : 0.70 0.80	0.80
0.79	*MM[0.95]N* tp : 0.80 0.89	0.90	0.76	*MM[0.9]N* tp : 0.76 0.85	0.86	0.70	*MM[0.8]N* tp : 0.69 0.78	0.80

Table 9. (Continued)

1-MWFI	D-MWFI	P-MWFI	RP-MWFI	MPPEL	1-MWFI	D-MWFI	P-MWFI	RP-MWFI	MPPEL
0.74	0.76	*UF[0.6/0.95]U* tp: 0.75 0.85	0.61	0.86	0.75	0.88	*UM[DEMIL]N* tp: 0.76 0.72	0.83	0.90
0.74	0.75	*UF[0.6/0.95]N* tp: 0.74 0.85	0.80	0.88	0.76	0.85	*MM[DEMIL]U* tp: 0.77 0.75	0.79	0.86
0.74	0.87	*UM[0.6/0.95]U* tp: 0.74 0.80	0.62	0.89	0.75	0.84	*MM[DEMIL]N* tp: 0.76 0.73	0.79	0.88
0.73	0.86	*UM[0.6/0.95]N* tp: 0.73 0.81	0.60	0.91	0.73	0.75	*UF[DESTR]U* tp: 0.72 0.74	0.72	0.78
0.74	0.80	*MM[0.6/0.95]U* tp: 0.73 0.83	0.61	0.87	0.72	0.74	*UF[DESTR]N* tp: 0.71 0.72	0.72	0.79
0.73	0.80	*MM[0.6/0.95]N* tp: 0.73 0.84	0.59	0.90	0.73	0.85	*UM[DESTR]U* tp: 0.72 0.75	0.72	0.90
0.77	0.80	*UF[DEMIL]U* tp: 0.77 0.76	0.78	0.80	0.71	0.82	*UM[DESTR]N* tp: 0.70 0.71	0.74	0.89
0.76	0.79	*UF[DEMIL]N* tp: 0.77 0.75	0.78	0.81	0.72	0.80	*MM[DESTR]U* tp: 0.72 0.73	0.72	0.84
0.76	0.90	*UM[DEMIL]U* tp: 0.77 0.75	0.81	0.91	0.71	0.78	*MM[DESTR]N* tp: 0.70 0.70	0.73	0.87

Table 10. Relative Performance of MPPEL and D-MWFI as a Function of Theoretical Error Incidence

	.95			.9			.8		
	D-MWFI Superior	equal	MPPEL Superior	D-MWFI Superior	equal	MPPEL Superior	D-MWFI Superior	equal	MPPEL Superior
M/S									
D1E1	2	2	2	1	2	3	1	2	3
D1E2	4	2	0	4	2	0	1	3	2
D1E3	3	2	1	3	2	1	3	2	1
D2E1	2	2	2	1	2	3	0	0	6
PFCD									
D1E1	0	4	2	0	3	3	0	3	3
D1E2	4	2	0	3	2	1	2	2	2
D1E3	2	4	0	2	2	2	2	2	2
D2E1	2	2	2	1	2	3	0	0	6

	.6/.95			DEMIL			DESTR		
	D-MWFI Superior	equal	MPPEL Superior	D-MWFI Superior	equal	MPPEL Superior	D-MWFI Superior	equal	MPPEL Superior
M/S									
D1E1	0	0	6	0	0	6	0	0	6
D1E2	0	0	6	3	0	3	0	0	6
D1E3	0	0	6	1	0	5	0	0	6
D2E1	0	0	6	0	1	5	0	0	6
PFCD									
D1E1	0	0	6	0	1	5	0	0	6
D1E2	0	0	6	2	2	2	1	1	4
D1E3	0	1	5	3	1	2	1	1	4
D2E1	0	1	5	1	1	4	0	0	6

3. Generally, on a case-by-case basis, the localization methods perform better on uniform error models than on nonuniform error models. Moreover, the relative superiority of MPPEL to D-MWFI is somewhat greater on uniform error models than on nonuniform models. Results are displayed in Tables 11 and 12.

Table 11 shows that MPPEL performance relative to D-MWFI generally improved on the nonuniform error models. If D-MWFI performed as well as MPPEL on the uniform error model and poorer on the nonuniform model, this is denoted as D-MWFI-to-MPPEL reversal. The converse is denoted MPPEL-to-D-MWFI reversal.

Table 12 presents evidence that on a method-by-method comparison, with only uniformity and nonuniformity of the error model varied, the success of the methods generally is relatively greater with the uniform error model. The exceptions seem to be MPPEL in terms of PFCD, where the greatest success is with nonuniform errors.

6. SUMMARY AND DISCUSSION

The effectiveness of various error-localization methods has been investigated in two different experimental designs (Appendix 1) and in assorted other simulations briefly reported in Appendix 2. These additional simulations investigated a variety of other MWFI methods and

Table 11. Relative Performance of D-MWFI and MPPEL Attributable to Uniformity/Nonuniformity of Errors

	D-MWFI-to-MPPEL Reversal	MPPEL-to-D-MWFI Reversal
M/S		
D1E1	7	0
D1E2	1	0
D1E3	3	3
D2E1	4	1
PFCD		
D1E1	7	2
D1E2	4	3
D1E3	7	1
D2E1	5	2

Table 12. Comparison of Performance on Uniform and Nonuniform Errors

	Superior Results Uniform Error			Equal Results			Superior Results Nonuniform Error		
	1-MWFI	D-MWFI	MPPEL	1-MWFI	D-MWFI	MPPEL	1-MWFI	D-MWFI	MPPEL
M/S									
D1E1	18	18	8	0	0	7	0	0	3
D1E2	15	18	12	0	0	4	3	0	2
D1E3	17	8	9	0	9	8	1	1	1
D2E1	17	16	7	0	1	7	1	1	4
PFCD									
D1E1	13	16	5	4	2	3	1	0	10
D1E2	10	15	12	8	2	5	0	1	1
D1E3	12	6	5	6	3	5	0	9	8
D2E1	16	10	4	2	5	4	0	3	10

the effect of different error rates in different fields. Results of these additional simulations were hardly conclusive; the methods could not be said to perform substantially better than those reported here. However, some evidence suggests that unequal error rates in fields lead to improved PFCD performance.

To test for representativeness of the results presented here, selected simulations were begun with different seeds, and the results were always comparable. The results presented in this chapter are believed to be representative of localization performance for low dimensional data spaces. The relative lack of differentiation among various weightings of MWFI could be attributed in part to the observation that for data spaces with dimension less than six, virtually all MWFI solutions have the same cardinality (Liepins, 1986). Localization experiments with higher dimensional spaces might show greater differences among the performance of localization methods than are reported here. Moreover, caution should be exercised in the extrapolation of these experimental results to practical applications; practical applications generally are high dimensional. Preliminary investigations (Liepins, 1986), suggest that, all other things being equal, localization performance deteriorates as the dimension (number of fields) increases. Also, it should be emphasized that the methods discussed in this chapter are applicable only to those records that are not individually important; the individually important records should be handled with a blend of these methods and rule-based methods. Given such caveats, the major conclusions of the experiments can be summarized in four points:

1. 1-MWFI performance appears to be relatively insensitive to error model. Variance in performance is smaller than the variance of the other methods studied. In practice, multiple 1-MWFI solutions should be expected. Such ties could be broken either randomly or with the use of an auxiliary discriminate function—D-MWFI.
2. D-MWFI nearly always performs better than 1-MWFI with ties broken randomly and, in the case of the independent error models, is thought to perform near the theoretical optimal. The drawback to D-MWFI is the additional computation required (in addition to the determination of all MWFI solutions). However, the computational requirements of D-MWFI are not nearly so severe as those of MPPEL (which is not advocated as a practical technique). Let n be the number of fields. Then, formally, MPPEL requires evaluation of each of the possible 2^n localizations. D-MWFI requires only as many additional evaluations as there are solutions to 1-MWFI.

3. If reliable information is available regarding the error model, it should be used in determining the weights of the MWFI (as in P-MWFI). However, it should be emphasized that refinement of the weights appears to be a double-edged sword—misspecification sometimes can lead to significantly inferior performance.
4. As the effective probability of error in each field approaches .2 to .3, the statistical error-localization techniques investigated in this chapter cannot be expected to provide satisfactory performance. Whether any method can deal adequately with such error-prone data remains an open question. Ideally, the experiments reported here represent the beginning of such investigations; controlled experiments with real data and with higher dimensional spaces are certainly in order.

A final caveat is in order. The methods investigated in this chapter address errors on a record–by–record basis. These techniques may be appropriate as a means to identify fields that are prone to error. However, as argued by Little in Chapter 10 of this book, record-by-record optimal localization followed by imputation may generate data that are unrealistically similar; the resulting distributions may be biased away from the true data. To preserve the true distribution, a different measure of optimality is required.

APPENDIX 1: EXPERIMENTAL DESIGN—DATA SETS AND CONSISTENCY REGIONS SIMULATED

A1.1 First Data Set

The first simulated data space D is a six-field example with the following fields:

$$D_1 = \{0, 1\} \qquad D_4 = \{0, 1, 2, 3\}$$
$$D_2 = \{0, 1, 2\} \qquad D_5 = \{0, 1, 2\}$$
$$D_3 = \{0, 1\} \qquad D_6 = \{0, 1, 2, 3\}$$

First Data Set Prior Distributions

The three prior distributions of true data were the restrictions of uniform, unimodal, and multimodal distributions to the consistency region as follows:

	uniform				unimodal				multimodal			
var 1	0.5	0.5			0.2	0.8			0.2	0.8		
var 2	0.33	0.33	0.33		0.2	0.2	0.6		0.2	0.2	0.6	
var 3	0.5	0.5			0.2	0.8			0.2	0.8		
var 4	0.25	0.25	0.25	0.25	0.1	0.1	0.2	0.6	0.4	0.1	0.1	0.4
var 5	0.33	0.33	0.33		0.2	0.2	0.6		0.2	0.2	0.6	
var 6	0.25	0.25	0.25	0.25	0.1	0.1	0.2	0.6	0.4	0.1	0.1	0.4

First Data Set Edits

Each of the three sets of consistency constraints was formulated in terms of explicit edits defined in Cartesian product normal form (Fellegi and Holt, 1976).

First Explicit Edit Set

$$
\begin{aligned}
e_1 &= D_1 \times \{0,1\} \times \{0\} \times D_4 \times \{0,1\} \times D_6 \\
e_2 &= \{1\} \times D_2 \times \{1\} \times \{0,1\} \times D_5 \times \{2,3\} \\
e_3 &= \{0\} \times \{1,2\} \times D_3 \times \{1,2,3\} \times D_5 \times D_6 \\
e_4 &= D_1 \times \{0,2\} \times D_3 \times D_4 \times D_5 \times \{0,1\} \\
e_5 &= \{1\} \times D_2 \times D_3 \times \{0\} \times \{1,2\} \times D_6
\end{aligned}
\qquad \text{(A1.2)}
$$

These edits specify that 422 of the 576 points in $D_1 \times D_2 \times \ldots \times D_6$ are unacceptable. For example, the record $y = \{1,0,0,2,1,3\}$ is unacceptable because $y \in e_1$. (This first edit set has been used previously in Liepins, Garfinkel, and Kunnathur, 1982; Garfinkel, Kunnather, and Liepins, 1985; and Liepins, 1984).

Second Explicit Edit Set

$$\begin{aligned} e_1 &= D_1 \times \{0,1\} \times \{0\} \times D_4 \times \{1\} \times D_6 \\ e_2 &= \{1\} \times D_2 \times \{1\} \times \{0,1\} \times D_5 \times \{2,3\} \\ e_3 &= \{0\} \times \{2\} \times D_3 \times \{1,2\} \times D_5 \times D_6 \\ e_4 &= D_1 \times \{0,2\} \times D_3 \times D_4 \times D_5 \times \{0\} \\ e_5 &= \{1\} \times D_2 \times D_3 \times \{0\} \times \{1,2\} \times D_6 \end{aligned} \tag{A1.3}$$

Inspection will reveal that only edits e_1, e_3, and e_4 differ from their counterparts of the first edit set and that these differences appear in the fifth field for e_1, second and fourth fields for e_3, and sixth field for e_4. However, even with these few changes, the number of unacceptable points is reduced to 245.

Third Explicit Edit Set

$$\begin{aligned} e_1 &= D_1 \times \{0,1\} \times \{0\} \times D_4 \times \{0,1\} \times \{1,2\} \\ e_2 &= \{1\} \times D_2 \times \{1\} \times \{0,1\} \times D_5 \times \{2,3\} \\ e_3 &= \{0\} \times \{1,2\} \times D_3 \times \{1,2,3\} \times \{2\} \times D_6 \\ e_4 &= D_1 \times \{0,2\} \times \{1\} \times \{2,3\} \times D_5 \times \{0,1\} \\ e_5 &= \{1\} \times D_2 \times \{0\} \times \{0\} \times \{1,2\} \times D_6 \end{aligned} \tag{A1.4}$$

Edits e_1, e_3, and e_5 differ from their counterparts of the first data edit set. The differences appear in the sixth field, the second field, the third and fourth fields, and the third field, respectively. The number of unacceptable points is further reduced to 212.

A1.2 Second Data Set

The second simulated example is a four-component data space with the following fields:

$$\begin{aligned} D_1 &= \{0,1,2,\ldots,27,28\} \qquad & D_3 &= \{0,1\} \\ D_2 &= \{0,1,2\} \qquad & D_4 &= \{0,1,2,3,4,5\} \end{aligned} \tag{A1.5}$$

This structure was chosen to examine the effect of having one field (here, D_1) with much larger cardinality than the other fields.

Second Data Set Prior Distribution

The three prior distributions of true data were restrictions of uniform, unimodal, and multimodal distributions to the consistency regions as follows:

	uniform			unimodal				multimodal			
var 1											
	each value 0.034			$x_1 = 13, 14,$ or 15			0.02	$x_1 = 4,$ or 14			0.02
				$x_1 = 13$ or 15			0.1	$x_1 = 4,$ or 14			0.23
				$x_1 = 14$			0.28				
var 2											
	0.333	0.333	0.333		0.2	0.2	0.2	0.4	0.2	0.4	
var 3											
	0.5	0.5			0.2	0.8		0.2	0.8		
var 4											
	0.167	0.167	0.167	0.12	0.12	0.12	0.12	0.3	0.1	0.1	0.1
	0.167	0.167	0.167	0.12	0.4			0.1	0.3		

Second Data Set Edits

The explicit edits in each case were:

$$\begin{aligned}
e_1 &= \{1,2,3,\ldots,28\} \times \{0\} \times D_3 \times \{0,1,2\} \\
e_2 &= \{16,17,\ldots,28\} \times D_2 \times \{1\} \times D_4 \\
e_3 &= \{0,1,2,\ldots,10\} \times \{0\} \times D_3 \times D_4 \\
e_4 &= \{1,2,3,\ldots,28\} \times \{1,2\} \times D_3 \times \{0\} \\
e_5 &= D_1 \times D_2 \times \{0\} \times \{3,4,5\}
\end{aligned} \tag{A1.6}$$

The edits specify that 749 of the 1044 points in $D_1 \times D_2 \times D_3 \times D_4$ are inconsistent. The edits were structured so that the proportion 749/1044 would be about the same as the proportion 422/576 in the sample of (A1.1)–(A1.2).

APPENDIX 2: ADDITIONAL SIMULATIONS

Methods that were investigated but were not part of the experimental design are reported here. Results were inconclusive, and only a brief selection is presented in this appendix.

Eight additional error-localization methods were investigated. Their definition depends on the equivalence of normal form and linear edits (Liepins, 1984).

DEFINITION 1 Consider a constraint $\Sigma m_j x_j \leqslant b_i$. Then, those fields D_j explicitly specified in the constraint are the active fields.

EXAMPLE Consider the constraint $2x_1 + 3x_3 + x_4 \leqslant 2$. Components 1, 3, and 4 are active.

DEFINITION 2 Consider the system of constraints $Mx \leqslant b$, and a failing record $\mathbf{y}$. Let $f_M(\mathbf{y}, i)$ be the number of times the ith component is active in failed constraints. Let $p_M(\mathbf{y}, i)$ be the number of times the ith field is active in satisfied constraints.

EXAMPLE Consider the system of constraints

$$\begin{aligned} &2x_1 + 3x_3 + x_4 \leqslant 2 \\ &x_1 + x_2 + x_3 \leqslant 3 \\ &x_1 + x_2 + x_4 \leqslant 4 \end{aligned}$$

and the record $\mathbf{y} = (1, 2, 1, 1)$. The record $\mathbf{y}$ fails the first two constraints and satisfies the third. Accordingly, $f_M(\mathbf{y}, 1) = 2$ and $p_M(\mathbf{y}, 1) = 1$, and similarly for the other fields. Two modifications of weights w_i of the MWFI can now be specified:

Modification 1 (Mod 1): $w_i = 1/(f_M(\mathbf{y}, i) + 1)$
Modification 2 (Mod 2): $w_i = (p_M(\mathbf{y}, i) + 1)/(f_M(\mathbf{y}, i) + 1)$

A heuristic motivation for these two modifications is, first, that the more often a field is active in failed constraints, the more it should be suspect; and, second, the more often it is active in satisfied constraints, the less it should be suspect.

The meaningful application of these modifications requires that the systems of constraints be expressed in such a fashion that the weighting is a function only of the acceptance region and not of the particular representation of that region. This is a technical requirement that can be satisfied by the reexpression of the constraints in a canonical form; see the discussion of maximal edits in Liepins (1980). Such a canonical representation was used in all simulations studied.

Table A2.1. Simulation Results for D1E1UF [ ]U, 1-MWFI, Mod-1, and Mod-2

Summary Statistics	Method	Error Model .95	.9	.8	.6	.4	.2	.1	.05
M/S	1-MWFI	.358	.296	.193	.066	.014	.001	.000	.000
	Mod-1	.360	.299	.197	.068	.014	.001	.000	.000
	Mod-2	.415	.346	.230	.075	.015	.001	.000	.000
PFCD	1-MWFI	.798	.771	.712	.588	.464	.334	.267	.206
	Mod-1	.798	.771	.713	.588	.461	.332	.266	.206
	Mod-2	.810	.782	.724	.596	.471	.341	.271	.206

Table A2.2. D1E1UF[ ]U with Large Spread Prior Errors

Simulation No.	Pr. Field 1 Error	Pr. Field 2 Errors	Pr. Field 3 Errors	Pr. Field 4 Errors	Pr. Field 5 Errors	Pr. Field 6 Errors	Average Pr. Error
1	.01	.20	.39	.01	.20	.39	.2
2	.11	.30	.49	.11	.30	.49	.3
3	.21	.40	.59	.21	.40	.59	.4
4	.31	.50	.69	.31	.50	.69	.5
5	.41	.60	.79	.41	.60	.79	.6
6	.61	.70	.89	.61	.70	.89	.7

Table A2.3. D1E1UF[ ]U with Medium Spread Prior Errors

Simulation No.	Pr. Field 1 Error	Pr. Field 2 Errors	Pr. Field 3 Errors	Pr. Field 4 Errors	Pr. Field 5 Errors	Pr. Field 6 Errors	Average Pr. Error
1	.1	.2	.3	.1	.2	.3	.2
2	.2	.3	.4	.2	.3	.4	.3
3	.3	.4	.5	.3	.4	.5	.4
4	.4	.5	.6	.4	.5	.6	.5
5	.5	.6	.7	.5	.6	.7	.6
6	.6	.7	.8	.6	.7	.8	.7

Table A2.4. Results of Simulations with Various Error Spreads for Mod-2 on D1E1UF, with Uniform Errors

Average Prob. Error	No Spread		Medium Spread		Large Spread	
	tp	PFCD	tp	PFCD	tp	PFCD
.2	.705	.724	.710	.743	.722	.768
.3	.630	.659	.642	.678	.653	.697
.4	.552	.596	.562	.612	.578	.632
.5	.470	.535	.482	.549	.495	.564
.6	.383	.471	.396	.484	.412	.500
.7	.292	.403	.307	.419	.322	.433

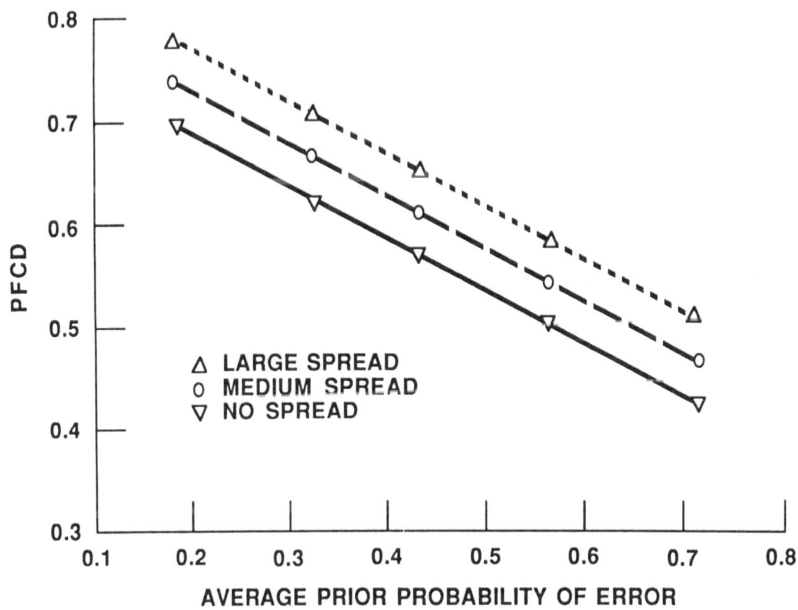

Figure A2.1. Simulation results for D1E1, average probability of error 0.2 for no spread, medium spread, large spread, with high error probability fields 3, 6, medium error probability fields 2, 5, low error probability, fields 1, 4.

Table A2.5. Variants of Large Spread Experimental Designs

Simulation No.	Pr. Field 1 Error	Pr. Field 2 Errors	Pr. Field 3 Errors	Pr. Field 4 Errors	Pr. Field 5 Errors	Pr. Field 6 Errors	Average
1	.39	.20	.01	.39	.20	.01	.2
2	.01	.39	.20	.01	.39	.20	.2
3	.20	.01	.39	.20	.01	.39	.2
4	.39	.01	.20	.39	.01	.20	.2
5	.20	.39	.01	.20	.39	.01	.2
6	.20	.01	.20	.39	.01	.39	.2

A third modification was MWFI with w_i defined by $\ln\{[f_M(\mathbf{y},i)+p_M(\mathbf{y},i)]/f_M(\mathbf{y},i)\}$ (taken to be an arbitrarily large number when $f_M(\mathbf{y},i)=0$). Modifications 4 through 8 were MWFI with w_i defined as the total number of edits multiplied by various powers of $p_M(\mathbf{y},i)/[f_M(\mathbf{y},i)+p_M(\mathbf{y},i)]$, the powers being 2, 1, 0.5, 0.25, and 0.125, respectively. The third through eighth modifications are not reported because they produced results only marginally different from those of Mod 2 on D1E1 and, in the limited simulations, none seemed different enough from 1-MWFI to warrant extensive reporting. The first set of results presented (Table A2.1) provides M/S and PFCD performance statistics for 1-MWFI, Mod 1, and Mod 2 for D1E1UF[ ]U, where the error model was varied from 0.95 to 0.05. The methods 1-MWFI and Mod 1 performed nearly identically; they were both dominated by method Mod 2.

Additional simulations were also made for Mod 2 with D1E1UF, independence of errors in fields, uniform distribution of errors, but with unequal prior probabilities of error in the fields. The two groups of designs are shown in Tables A2.2 and A2.3 which give the experimental designs for large and medium spreads in errors, respectively. Table A2.4 presents the PFCD performance results for these designs.

Figure A2.1 presents the graphs for the data of Table A2.4. Tables A2.5 and A2.6 presents evidence that even with the error spread and average error held constant, PFCD performance varies substantially as different combinations of fields are chosen to have a high incidence of errors.

Table A2.6. Large Spread Results for Mod-2, D1E1UF[ ]U Showing Substantial Differences in Performance as a Function of What Fields Are Most Likely to be in Error

High (.39) Prob. Error Pair	Medium (.20) Prob. Error Pair	Low (.01) Prob. Error Pair	tp	PFCD
3, 6	2, 5	1, 4	.722	.768
1, 4	3, 6	2, 5	.692	.706
3, 6	1, 4	2, 5	.708	.733
2, 5	3, 6	1, 4	.715	.756
1, 4	2, 5	3, 6	.692	.700
2, 5	1, 4	3, 6	.708	716

REFERENCES

F. M. Andrews and A. R. Herzog (1986). The Quality of Data as Related to Age of Respondent, *J of the American Statistical Association, 81, no. 384*: 403–410.

I. P. Fellegi and D. Holt (1976). A Systematic Approach to Automatic Edit and Imputation, *J of the American Statistical Association, 71, no. 353*: 17–35.

R. J. Freund and H. O. Hartely (1967). A Procedure for Automatic Data Editing, *J of the American Statistical Association, 62, no. 318*: 341–352.

R. S. Garfinkel, A. S. Kunnathur, and G. E. Liepins (1988). Error Localization of Erroneous Data: Continuous Data, Linear Edits, *SIAM J of Scientific and Statistical Computing, 9, no. 5*: 922–931.

R. S. Garfinkel, R. S. Kunnathur, and G. E. Liepins (1986). Optimal Imputation of Erroneous Data: Categorical Data, General Edits, *Operations Research, 34, no. 5*: 744–751.

M. E. Gonzalez et al. (1975). Standards for Discussion and Presentation of Errors in Survey and Census Data, *J of the American Statistical Association, 70, no. 351, Part II*: 5–23.

B. Greenberg and R. Surdi (1984). A Flexible and Interactive Edit and Imputation System for Ratio Edits, *Proceedings of the Section on Survey Research Methods, ASA*.

F. T. Juster (1986). Response Errors in Measurement of Time Use, *J of the American Statistical Association, 81, no. 394*: 390–402.

G. E. Liepins (1986). Unpublished results, Oak Ridge National Laboratory, Oak Ridge, Tennessee.

G. E. Liepins (1984). *Algorithms for Error Localization of Discrete Data*, ORNL/TM-8183, Oak Ridge National Laboratory, Oak Ridge, Tennessee.

G. E. Liepins (1980). *A Rigorous Systematic Approach to Automatic Data Editing and its Statistical Basis*, ORNL/TM-7126, Oak Ridge National Laboratory, Oak Ridge, Tennessee.

G. E. Liepins, R. S. Garfinkel, and A. S. Kunnathur (1982). Error localization for Erroneous Data: A Survey, *TIMS/Studies in the Management Sciences, 19*, North Holland Publishing Company, pp. 205–219.

R. J. A. Little and D. B. Rubin (1987). *Statistical Analysis with Missing Data*, John Wiley & Sons, New York.

W. G. Madow, I. Okin, H. Nisselson, and D. B. Rubin (1983). *Incomplete Data in Sample Surveys*, Vol. 1–3, Academic Press, New York.

J. I. Naus (1975). *Data Quality Control and Editing*, Marcel Dekker, New York.

J. I. Naus, T. G. Johnson, and R. Montalvo (1972). A Probablistic Model for Identifying Errors in Data Editing, *J of the American Statistical Association, 67, no. 340*: 843–850.

L. Pritzker, J. Ogus, and M. H. Hansen (1965). Computer Editing Methods—Some Applications and Results, *International Statistical Review, 41*: 442–472.

D. B. Rubin (1987). *Multiple Imputation for Nonresponse in Surveys*, John Wiley & Sons, New York.

Sande (1976). Unpublished. Statistics Canada.

B. D. Spencer (1985). Optimal Data Quality, *J of the American Statistical Association, 80, no. 391*: 564–573.

W. J. Stuart (1965). Computer Editing of Survey Data—Five Years of Experience in BLS Manpower, *J of the American Statistical Association, 61*: 375–383.

M. E. Terry (1963). The Principles of Statistical Analysis Using Large Electronic Computers, *Bulletin of the International Statistical Institute, 40*: 547–552.

W. H. Williams (1978). How Bad Can "Good" Data Really Be? *The American Statistician, 32, no. 2*: 61–65.

T. Wright (ed.) (1983). *Statistical Methods and the Improvement of Data Quality*, Academic Press, New York.

Index